Creating Positive Change

AF797772

Creating Positive Change

Creating Positive Change

Nicholas Gambacini

Right House Publications™

Right House Publications

Creating Positive Change

A Book Where You Are A Creator • The Art of Positive Change • A Tribute To Humanity

CREATING POSITIVE CHANGE ©2018.
Second Edition.
Published in 2019.

First Published: September 5th, 2018

Editor: Emily Sirko

ISBN: 978-0-692-18557-5

"Creating Positive Change" is an official elected Member of the ***Finnerman's Authors Association***, and their Charitable Donors Cause. Further details in back.

Affiliate of *Ohmmissions Productions* ™

All Rights Reserved.

No part of this book may be used or reproduced in any manner whatsoever without the written permission of the publisher. Printed in the United States of America.

Books available for special promotions and premiums. Contact Right House Publications for information:

P.O. Box 701 Northville, MI 48167 Or
Righthousepublications@gmail.com

Other Works by this Author:

The Petrichor Series: The Last Standing Army, The Final Stand, and The Awakening

Foreword

Hello reader!

This book is purposed with healing negativity. As a guide, and a reference, Creating Positive Change will take you through universal principles in an easy-to-understand format. By bridging gaps between Neuroscience, Psychology, Physics, Nature, Energy, and more, enjoy the adventure as you explore the wonderful world of Positivity, and all its incredible power.

—Carpe Diem!

INDEX:

(Cont.)

1

Oneself and The Railway Model

"The laughter of children is so pure for a reason."

(I)

Introductory Topics

Positivity Begins with Purging, and Purging is an elaborate and systematic way of cleansing oneself of negativity. Purging is the first step towards creating Positive Change, and Purging requires a preliminary understanding of the concepts outlined within this section. Before reading on, and in order to understand these concepts, one should first begin to understand and establish a concept of what ***you****, (hereafter referred to as "oneself"), considers "oneself" to be. **NOTE: This may involve defining one's own physical and mental limitations, which are usually subconsciously outlined by one's thoughts that begin with "I...", and in the self-defeating sense "I cannot..."*

The Comfort Zone and Understanding Oneself

One's idea of "oneself" can be created and limited by a few factors, of which includes one's individual **Comfort Zone***. One's Comfort Zone is defined by the structure or limitations one sets upon oneself, which is commonly known as one's "image". In the era of social media, we see more and more of this concept of "image" and "oneself" strongly emerging in all generations and ages.*

A Note Before Entering The Realm of Positivity

Let us not forget to smile in the process of Creating Positive Change, knowing that you have already begun a first step towards cleaning up any negativity in your own life! 3... 2... 1... and breathe!

Food For Thought

In the ***Concept***[1] *of oneself, lies the fun of opening one's mind. There is an infinite amount of conceivably possible definitions for "oneself", due to the fact that there is an infinite amount of possible "oneselves" conceivable by human imagination. Positive Change is only accomplishable from an* ***Empath***[2]*, and a true Empath is someone who can understand that there is an infinite amount of possibilities of "oneself" that all individuals are capable of. This understanding allows for emotional intelligence, which will be discussed throughout this literary work, from multiple perspectives.*

[1] *A* ***Concept*** *is a compilation of thoughts that create a completed idea. Concepts vary in strength by the amount and legitimacy of the supporting thoughts creating it.*
[2] *An* ***Empath*** *is able to look at another human and understand exactly what they consider to be their individual "oneself".*

And now, let us begin a journey of positive understanding, starting with...

The Camera Analogy

The Camera Analogy *is a useful tool to understand one's operation within the universe, and one's roles and abilities within a given System that one is interacting with. It will be set in progressively refined lights throughout Creating Positive Change. In The Camera Analogy, one is to think of oneself as a camera, and a System as what one sees through the camera* **Scope**. *Just as a camera has the ability to zoom in and out, far and wide, so too does one have the ability to see any System that falls within, but not outside of, one's adjustable camera Scope for the given moment that one is making an observation.*

Patience

Patience[3] *is the practice of waiting to ensure a System is fully understood before one acts upon it, and a* **System** *is all the connected entities of anything that exists in the Physical Realm. Patience is an open acceptance of the fact that things are not always what they seem on the surface, due to the infinite amount of overlapping Systems that complete a "Stream"*[4] *of energy. Without Patience, any of those deeper, "under the surface" concepts may be*

[3] *It is important to remember that the practice of* ***Patience*** *entails one will have no expectation of a deadline, and therefore time is not to be considered a limiting or set factor. Since any given process is completely unique and independent, it will therefore have a unique time allotment and process completion time.*

[4] *The "Stream Analogy" is explained in Section (IV) "Can a Stream Flow with a Dam in its Way?"*

overlooked (the out of Scope views), and the System will not be fully understood. This leads to assumptions, and judgment.

Patience entails that one can only look at the world through a camera Scope, or no camera Scope, meaning there is a limit to the amount of information one can process for a given moment, and the information that one can process depends on what information one was exposed to. Patience tells us that things are not always what they seem, due to the different pictures taken by the camera, or what is happening outside the Scope that one doesn't see, which affects the Systems within the Scope. To further elaborate, Patience is a practice of preventing any one of those out-of-camera views from being misunderstood, by acknowledging that the entire picture may sometimes be a combination of all types of camera shots: close ups, far away shots, wide angle shots, shots one can't see or hasn't taken yet, and more! In a tangible application, a lack of patience can lead to one's own improper assumptions and misjudgments.

Food For Thought

When trying to create Positive Change, Assumptions and Judgement lead to most all possible complications, and the camera concept is an excellent way of understanding how Systems interact, and how one's individual perspective is formed in a given moment. Now, let us Patiently move forward— to Frequencies...

(II)

The Building Blocks of Creating Positive Change

In this section, we will be exemplifying key concepts and major tools that will be the basis for the remainder of this literary work. Delve in with an open perspective on these topics, through the filter of how to utilize them for Creating Positive Change in your own individual journey! Additionally, practice Patience throughout this section by taking your time to read with a gentle kindness to oneself. Some of these concepts may seem complicated, but they will be further explained, and they will be the key to creating your own ease of understanding of how to Create Positive Change for yourself perpetually!

Frequencies

You may have a concept of Frequencies already, so here is where you must challenge yourself to somewhat forget that concept. Open yourself to a new idea of Frequencies: as if it is the first time you have ever heard the word in your entire life, and you are hungry to learn more! Later, you may combine the two concepts (your old idea and this new one) in order to create a more complete and refined understanding, but to begin, and in true learning, one must require oneself to become an unbiased "tabula rasa"[5]*, a phrase that was made popular by Behavioral Psychologist B. F. Skinner. That being said, allow us to view the role that Frequencies play in Positive Change.*

[5] *Check Glossary for* ***Little Albert Experiment****.*

What are Frequencies?

Frequencies are understood to be a particular "vibration" pattern. For purposes of this book, Frequencies and vibration will be interlaced with the general word "energy". There is a basic understanding uniting Energy, Frequencies, and Vibration: that energy is, at its foundation, perceived to be in multiple positions at once (or moving), such as with an electron in an electron cloud, therefore it is commonly thought of as "vibrating" (or moving) at a given Frequency pattern (or energy pattern).

The Relationship of Energy and Frequencies

Now that we have brought to light that energy is seen as the glue that foundationally holds "you" all together, and that it is known to be moving (or vibrating), we can summarize that "you", a fleshy collection of different energies, are composed of ***Frequencies****: the resulting vibrations of the energies of a particular System* *(you), at the time of which it (you) is observed. Again, in this work, energy and vibration are both terms that, on a foundational level, describe the same entity.*

"Oneself"

With this idea of Frequencies established, we can now safely assume that, in solely the Physical Realm, one has an electronic field of energies shifting and moving between what one has defined "oneself" to be, and the surrounding Frequencies outside of "oneself". This interconnectedness and interacting ability completes the basis of understanding for Frequencies.

***REMINDER: Again, at any given point, a particular System (let's say you, for example) is defined by the Frequency at which it (you) is vibrating at the time of which it (you) is observed.*

The "Railway Model" and a Central Dogma

The Railway Model[6] *is a hypothetical model that illustrates the observable truth that one's Systems have Frequencies that are interacting with environmental Frequencies, much how a Railway System operates with multiple entities (railroad stations) through certain channels (railroad tracks). We will utilize 'the Railway Model' as a simplified version of what physicists describe as* **String Theory**[7]*, and we will do this*

---

[6] *The* ***Railway Model*** *is a tool used to simplify and describe the flow of energy between "oneself", one's Frequencies, and the environment, and highlights their interconnectedness. A Proper flowing Railway Model is key to deeming a successfully completed Purge!*

[7] ***String Theory*** *is the connection of all matter and concepts to one another in the Physical Realm. It is a vast topic, but in a general sense, it means everything is connected in some way, like strings of varying strength. The example of the Stream, in Section (IV) "Can a Stream Flow with a Dam in its Way?" will help to elucidate upon the interconnectedness of things of matter.*

by highlighting a central dogma of the Railway Model.

The Central Dogma of the Railway Model and Further Explanation

The two Systems that have become our primary focus – "oneself", that which views and takes the 'in-Scope' photos, and the environmental Frequencies (or the frequencies outside of the camera Scope) – are directly interconnected through a series of ongoing energetic processes, at all times, regardless if one is aware of it, or chooses to acknowledge it. In the case of humans, breathing is an example of one such link to environmental Frequencies. Breathing is an exchange of physical molecules from the environment (one station), through capillaries in the lungs (Railroad Tracks), into the bloodstream (Railroad tracks), and to the necessary organs (another station). The Railway model also illustrates that connections in a

System are typically numerous (i.e. the many blood vessels in the human body), seen by perspective (the in-Scope camera shots), and are an open network of entities (they are all connected in some way, if even not directly).

***NOTE: For emphasis, and in The Camera Analogy, environmental Frequencies are the Frequencies outside of the camera Scope, meaning anything outside of one's known knowledge.*

In the case of individuals, the Railway model can be used to highlight that due to the fact there is a network of connections between oneself, and the environmental Frequencies, one can safely assume that the energy Systems that comprise oneself are in need of a ***Proper***[8]

[8] *For the purposes of this book, the word* ***Proper*** *is used to describe a well-maintained connection, something*

surrounding environment (and functionality) in order to create a Proper Railway Model. For example, can a train travel if the railway is rusty, old, and neglected? No, or at least not effectively. Dually, energy cannot travel through a System effectively under the same hypothetical hinderances, such as a neglected connection!

The importance of a proper environment is conversely exemplified with a ***Toxic***[9] *environment. A Toxic environment will work to create a Negative flow of energy (since oneself will be interconnected with Toxins), and will be later elaborated upon.*

Respected (defined on Pg. 29), or in the occasion of a Proper environment, an environment in which Positive Change can flourish.

[9] ***Toxins*** *are defined as people and things that have a Negative or Imbalancing influence. This will be later discussed in greater depth.*

Food For Thought

One will become the very Frequencies one is submerged in.

Respecting One's Railway Model

Inherent control of a Proper Railway Model requires that one utilizes ***Respect***[10]*, and in order to understand how to Respect one's Railway Model, one must first understand how to Respect oneself. The practice of Self-Respect is defined best by famous psychologist, Abraham Maslow, in a concise and pyramidal design known as "Maslow's Hierarchy of Needs".*

Maslow's Hierarchy of needs defines ***Respecting***[11] *oneself by establishing the essential*

[10] ***Respect*** *is the practice of understanding the power and ability something has.*

[11] ***Respecting*** *is the act of giving Energy to something in order for its Positive development. One can utilize Patience for the best results.*

duties every human must maintain in order to sustain a baseline of existence or more as the bottom level of the pyramid.

Maslow's Hierarchy of Needs

At the base of the model pyramid, exists the absolutely essential forms of Respect that must be present in order for a Physical ***Being***[12] *of matter to exist in the Physical Realm, while the upper levels of the pyramid outline everything else (mainly the needs of one's Higher Self, or the self that exists for more than the baseline human essentials).*

It is primarily the baseline needs (food, water, and sleep) that determine how one's Railway Model is functioning; therefore, they will be the primary example throughout this literary work.

---

[12] A ***Being*** *is classified as something that has an ability to create change in their environment through manipulation of a Physical form, involving a form of consciousness.*

Again, these three baseline needs are: food, water, and sleep, and the quality of each of these entities will play a role in the Railway Model's ability to function.

Food For Thought

As an obvious restatement, by Properly acknowledging and Respecting the needs of one's daily life, one allows for the Proper control of vibrations through The Body. Thus, it is these three essentials that allow for Maintenance of one's Railway Model, and furthermore it is the purity of these three essential needs that control how well the Railway Model functions. After all, "you" are what "you" eat.

**Notice how materialism was not listed in this baseline of necessary essentials of existence?*

The Railway Model (Further)

The Railway model expands beyond just what is observable through the Scope (the Physical Realm) and that outside of the Scope (the unseen and unknown aspects of the Physical Realm that have an effect on that which is in the Scope) to a third entity: the entity that utilizes the camera itself (that which is doing the viewing). This entity, for purposes of Positive Change, is known as the ***Higher Self****, and is above the camera and the Scope in terms of existence. For now, we can conclude the concept of the Railway Model by understanding it encompasses the interconnectedness between oneself (the self beyond the Physical Realm) and the Physical Realm (the observable and unobservable universe) including specifically* ***The Body***[13]

[13] ***The Body*** *is the conduit between thoughts of the Higher Realms and the Manifestation of those thoughts in the Physical Realm. It is one's link to the Physical Realm, and*

(the link to the Physical Realm). The Railway Model is supported by a few genres of information, including Science, which considers the crossroads of these connections to be a physical location in the body (known as the brain: an electronically charged organ), as well as Sanskrit, which defines it as the "Chakras" (the metaphysical energy centers), and so on. All these concepts have a basis in energy.

***NOTE: As one further refines how The Camera Analogy directly applies to oneself, The Body can now replace "the camera" as the tool which creates an observable Scope of our present Physical universe.*

Positive and Negative Energy

For the purposes of this book, ***Positive Energy (Positivity)*** *will be thought of as any energy or*

---

the defining aspect in the ability for one to have a Railway Model.

combination of energies that support a Proper Railway Model. Positivity is known to lead to ***Feelings***[12] *such as happiness, liveliness, and inspiration, and are contrasting forces to Negativity.*

Negative Energy (Negativity) *will conversely be thought of as any energy or any combination of energies that support a Negative environment and an improper Railway Model. Negativity can lead to depression, destruction, or a lack of enjoyment.*

***NOTE: To truly understand the difference between Positive and Negative Energy, one should first successfully complete a Purge, and*

[12] ***Feeling*** *is the process of experiencing a Frequency. When one experiences a feeling, one is given a choice: to Manifest that particular Frequency, to "tune in" to another Frequency, or to do nothing. In order to effectively choose, one must practice Discipline (defined on Pg. 72).*

the process of Purging is thoroughly outlined in Chapter 2.

Food For Thought

Positive people are creators, and Negative people are destroyers.

Recap, the Physical Self, and an Introduction to the Basic Principle of Humanity

At this point, it is now understood that one's Physical Self is one's link to the surrounding environmental Frequencies (as aforementioned in the Railway Model), and that the Physical Self is commonly referred to as, The Body. It is also understood that one can help or hinder the proficiency of one's ability to ***Manifest***[14] *one's*

[14] *In terms of Creating Positive Change, the act of* ***Manifestation*** *is defined as the transference of thoughts, dreams, desires, etc. (ultimately, the Positivity one wishes to create) from the Higher Realms into the Physical Realm.*

thoughts into the Physical Realm, by two major factors: one) by choosing how one treats one's Railway Model, and two) by how well one maintains the greater connection System (that inside of the Scope and outside of the Scope, as well as the Body- the conduit for which energy flows from the Higher Self, to the physical self, and outward into the Physical Realm).

Furthermore, the understanding that the control of one's Physical Self has been bestowed to each of us (including you, the reader of this work) in some way (however one believes—through fate, God, the Universe, spontaneity, just plain no reason, etc.), signifies that this ability of self-control is a universal truth for all humans. In simpler terms, everyone has ownership over their own body, and therefore their own journey, regardless of the condition of the particular Body one is bestowed ownership over, or how efficient they are with that particular control.

This truth will hereafter be referred to as ***The Basic Principle of Humanity****, and grants the ability of choice, and signifies consciousness.*

The Basic Principle of Humanity (BPH)

The Basic Principle of Humanity (or BPH) entails that by Respecting oneself (according to the baseline of Maslow's Hierarchy), one is also Respecting others, and will most likely not be projecting a Negative Frequency. This principle also entails that since, as Physical Beings, we are all a part of the same System (the Physical Realm and its corresponding fundamental units of energy), we therefore can be considered separate entities of the same whole. This elaboration proves that by helping another Being, one is in fact helping oneself, and vice-versa, due to the Physical Realm's universal interconnectedness (String Theory, as exemplified by the Railway Model). This holds true for Positive and Negative people.

In creating Positive Change, it is important to note that one should not concern oneself with attempting to help all Beings, as some Beings are at different points in their individual journeys, and it is not one's responsibility to grow and develop everyone, just to aid and supply them with any Respect that one can spare, without disrespecting oneself first. After all, we (humanity) are only as strong as our weakest link.

***NOTE: This connectedness to each other is also why when one hurts or helps another human being, one is capable of understanding the emotions of that human being (known as empathy, and backed in psychology by the idea of "mirror neurons"; also explaining the possibility of "tuning in" to another's Frequencies). There are many known theories supporting the connectedness of humans, therefore it is a universally accepted truth used for the purpose of strengthening the concepts of*

Creating Positive Change. Furthermore, since humans are all comprised of the same universal energy (atoms at a foundational level), it can be understood that this foundational similarity unites us, and that the major difference between individuals (on a greater level than the superficial appearance) is the observable Frequencies to which they are tuned in to, at the time of which they are observed. Again, these frequencies are determined by one's thoughts, and allowing one's thoughts to tune in to or align with another's would align the frequencies and allows for team communication on any level to occur.

Food For Thought

As a general rule, if the System as a whole functions Properly, then every component functions Properly, and if every component functions Properly, then the System as a whole functions Properly. A car, for example, cannot function efficiently if a wheel is broken or the engine doesn't work, and so too is the same true of the Universe, and as aforementioned, the Railway Model. Again, the concepts mirror each other, and are connected by String Theory. Respect works to Maintain a functional balance in all entities of the Physical Realm.

Depression

As defined in this book, **Depression** *is a state of absolute Negativity. One that is depressed may feel as though they are "trapped", "limited", "unable to..." or any other statements of this sort. Fortunately, if one is depressed, one need only understand that it is likely due to a reversible overflow of Negative Energy, or a disrespected Railway Model! Both causations* *are able to be resolved, which will be explained further in the main application of a Purge.*

Toxins Re-Established

The topic of Toxins is hugely important, and therefore will be set in progressively refined lights as we advance through ***Creating Positive Change.*** *As aforementioned, Toxins are defined as people and things that have a Negative or Imbalancing influence and can result in a hindrance of one's ability to have a Proper functioning Railway Model. Without a Proper functioning Railway Model, one becomes "lost", "out of touch with oneself", and eventually Depressed since the Body can no longer connect the Higher Self to the present Physical Realm. These are feelings which all stem from the improper treatment of one's Body.*

Facilitated Learning Between Humans

As ***Beings of Consciousness***[15] *we have the ability of* ***Facilitated Learning****, or the purposeful passing of information from one Being to another. Facilitated Learning can take many forms, some of which include, but are not limited to, tactile, auditory, and visual segues of information transfer. Facilitated Learning will be further developed in later chapters, but for now, the baseline implication of Facilitated Learning is that information can be successfully passed and absorbed between Beings under the correct circumstances, and is determined successful if a Being can Manifest the concept on one's own, without any outward assistance.*

[15] *At this point, understand that humans are capable of analytical thought and organized energy conduction, giving us "consciousness". Consciousness grants humans the ability of awareness, and therefore the ability of creating Positive Change, with Proper understanding.*

Food For Thought

Learning is facilitated, not forced.

A Brief Introduction to the Realms

Since there are many known conceptions of Realms, in this literary work we will keep things simple by classifying the known Universe (that inside and outside the camera Scope, as well as the camera – aka The Body) into one of two categories: ***The Physical Realm****, or* ***The Higher Realms****. These Realms can be broken down as such:*

1. ***The Physical Realm***

   The Physical, the tangible, things with mass, and things with measurable energy loosely describes the Physical Realm. In a further explanation, the Physical Realms encompass that which is limited by the Physical laws of Nature, and bound by fixed points in Time. In The Camera Analogy, the

Physical Realm includes the camera (The Body), that which falls within the camera Scope, and that which falls outside of the camera Scope.

2. **The Higher Realms**

   One's Being, one's Higher Self; the visual and limitless Realms (not limited by the laws of the Physical Realm), and the origin of the higher forces (all possible Systems of the Physical Realm) all comprise the Higher Realms. Sometimes this realm is referred to as the Realm of Creation. In The Camera Analogy, the Higher Realm is that which controls the camera, and is understood to be a realm inconceivable by human conception, due to the fact that the realm itself exists outside of the Physical Realm. In this occasion, the word 'outside' is considered a loose term, since words themselves are limited by the laws of the Physical Realm as well, and would fail to be able to explain the Higher Realms (which are not bound by time).

***NOTE: Again, for purposes of simplicity, one can classify everything into either the Physical Realm (The Body, and things bound by fixed points in Time), or the Higher Realms (thoughts, and everything else).*

(III)

The Higher Self and Further Into Beings

The Higher Self is the "oneself" in this whole idea. It's you! It's the thing (the Being) that experiences it all: the feelings, the thoughts, the thing that thinks, the thing that becomes self-conscious, and the thing that creates. "You" exist on multiple planes of existence, and it's time to learn how to utilize these Realms as a key to unlocking the ability of Positive Change.

The Physical Realm, and a Further Introduction to the Higher Realms

Since thoughts do consume energy of the Physical Realm (i.e. one can become tired from thinking too much), we can assume that there is a link between thoughts, and the energy that comprises the Physical Realm. The Physical Self (The Body) fosters the connection to one's thoughts (which do not exist in the Physical Realm, for they have no mass, and currently no known quantifiable Physical location) and to the Higher Self (the "you": the thing that makes the thoughts). This connection and the strength of this connection follows the dogma of the Railway Model.

For example, if one were to verbally repeat the words, "I am happy" (an action and Manifestation in the Physical Realm), enough times, and if one were also able to truly allow oneself to **Believe**[16] *it by letting go of all other energies (other thoughts), then one will find that one is solely* **Focused**[17] *on the thought of being happy, and the Physical Manifestation will be a happy Physical Self. This ability to create any Positive process stems from the Higher Self, and from smoothly guiding the Stream of thoughts from the Higher Self to the Physical Manifestation. The efficacy of this process of smoothly guiding a Stream of thoughts from the Higher Self to the Physical Manifestation is reliant on two major factors: the awareness one*

[16] *See Glossary for "***Believing***"*

[17] *The act of* **Focusing** *involves choosing one entity to think about, and Maintaining a directed thought flow towards conceptions of that one entity, for any duration of time, in order to Manifest it. Focus is a method of controlling energy flow.*

has of the process, and the Properness of one's Railway Model.

Food For Thought

When more information is accrued about a particular belief (Learning), refinement of that particular Belief occurs, sometimes changing it entirely. Conversely, when a Being is not open to refining a concept, they can be deemed "stuck", or considered to be in a state known as "blocked" (a negative state in which they are unable to grow and develop further). But when a Being believes they do not yet know everything, and are therefore always open to learning and changing, then that Being becomes limitless in terms of where they can grow and develop. Since they will always be under the impression that they do not know everything, limitless Positive possibilities open in terms of how much they can grow!

Understanding Beauty for Positive Change

Beauty *can be seen in two ways: Beauty that is created, and Beauty that is found naturally occurring, but in either occasion, Beauty is* *always the observable result of Positive Energy. If one is miserable and is longing for a better, more beautiful life, then one should Focus only on Positive Energy, or Positive Frequencies, and therefore one will find, or Manifest, Beauty. This is known as "The Law of Attraction" – later explained in Chapter 4.*

*In the Beauty that is created, one can quantify Beauty to be the result of a maximum amount of Positive Work put in for a given unit of Time (*further explained in Chapter 3 – "The Results Equation"), and is the result of an input of only or mostly Positive Energy (which explains the sensation of warmth or light when experiencing an entity of Beauty).*

For comparison purposes, it is understood that, ***1)*** *the Beauty that is created stems solely from the beholder's thoughts, and* ***2)*** *Naturally Occurring Beauty differs from Created Beauty in that it is created by a source deemed unknown to the beholder. Furthermore, in the Beauty that* *is naturally occurring, Beauty is in the eye of the beholder, since every Being is different, and resonates at different Frequencies. It is as the saying goes: There are many roads up the mountain, meaning two entities can find different ways to the same observation, even if that observation is different for each individual's perspective (depending on where their camera Scope is focused).*

Food For Thought

As aforementioned, every Being is different, and Beauty is in the eye of the beholder, but this does not rule out the possibility of two or more Beings deeming the same entity to be beautiful (in their own unique perception). Instead, it entails that, between the two Beings, although the same end result occurs (the deeming of something to be beautiful), there are different events that may have led up to this same result. This is a tremendously important acknowledgment: that others may see an entity differently or the same, and with this acknowledgment, one can effectively understand how to express oneself to others, and the inherent need of understanding where others may come from in their own understanding of the same particular entity.

Now, allow us to look at The Stream Analogy...

(IV)

Can a Stream Flow With a Dam in its Way?

Can a Stream freely Flow with a Dam in its way? No! Therefore, energy cannot flow freely through a System if it is blocked in any way. But how did the analogical Stream become blocked to begin with? Let us observe The Stream Analogy, how it relates to the concepts of the Railway Model, and how it is the key to understanding the implementation of 'Creating Positive Change's' major concepts.

The Stream Analogy

The Stream Analogy *is used to simplify a greater concept of how the flow of energy in the Physical Realm operates (further expanding on String Theory and The Railway Model), and to also illustrate how one can self-correct broken or blocked flows. In The Stream Analogy, one is presumed to be measuring a hypothetical Stream at a point in which it is blocked. Why it is blocked is undetermined at the time of the observation. Perhaps over time debris gathered, and blocked passage. Perhaps old farmer Yeller and his dog Skip were bought out by a big corporation, and had to move far away where they could no longer Maintain the hypothetical Stream on their Property. Perhaps a bad storm knocked a tree in the Stream's path— but in the infinite amount of possibilities of why the Stream is no longer flowing, there is a distinct consistency: the transformation of the Stream's form from flowing, before it was blocked, to the* *state that it is at the time of the observation, known as blocked. This result exemplifies an Imbalance in the Proper upkeep of the Stream's natural state (known to be flowing). Without a flowing Stream, there is no Stream at all. The observation that the Stream had once been flowing and is now blocked leads one to a second important dogma in Creating Positive Change: that Imbalances create blockages.*

***NOTE: This dogma is key to understanding the concept of Purging, and to being able to fully explicate its Proper use in all areas of one's life, as well as to utilize the tools of Creating Positive Change.*

The Purpose of a Purge

In order to fully understand the purpose of a Purge, one needs to understand how to prevent Imbalances from occurring (or, if they are already happening, one needs to understand how to prevent the Imbalances from getting any worse, with the goal of eventually strengthening one's overall System, or one's Railway Model). Thus, the Stream analogy is our first step in understanding how to successfully complete a Purge. Again, a Purge is the first step in successfully completing Positive change on any scale, as well as preventing future Imbalances. The purpose and how-to of Purging will be later explained in greater depth, but for now it will only need be understood as a process that takes something that is "blocked", "broken", "hopeless", "without direction", or "Negative" and brings it back to its natural flow.

Imbalances

Imbalances *create blockages in a naturally flowing System, which will affect the larger System as a whole (The Railway Model). When one's Railway Model is not balanced, one will impact not only one's own Systems, but the*

Systems that one is surrounded by, and connected to (String Theory). To further elucidate this point, entertain the possibility of a literal Stream being blocked, and the consequent outcomes:

1) The Stream becomes blocked;

2) Fish "DownStream" start perishing because they no longer have a Stream to swim in;

3) Fishermen are out of work because they no longer have fish to catch;

4) Local food shops then go out of business with a lack of supply;

5) Local town hunger rates increase because of a shortage of their primary food supply (fish);

6) Townsfolk desert their homes to find jobs elsewhere (since most worked at the local food stores) as well as to search for food; and finally,

7) Home property values decrease because of desertion.

The result of decreasing property values due to the Stream being blocked stems from a small Imbalance in a much larger system, and could almost be thought of as unrelated, because it is so far ***DownStream***[18]*, or down the connection cycle (things outside of the camera Scope), that one might not initially notice any connection at all. What does this mean for "you", and Positive Change? It means one must understand that any one System will always have resounding effects on its linked, and perhaps initially thought to be unrelated, Systems. For emphasis, in the Physical Realm, one can always assume that a System is in fact linked to other Systems somehow, in some way (String Theory) and the effects of that link can be defined and understood by the Railway Model, and the Results Equation.*

[18] ***DownStream*** *represents the resulting effects of one or more Imbalances that can be traced a few links back in a connection cycle.*

Quick Self-Reflection on Imbalances

1. *What Systems do you feel are Imbalanced in your life, and could be causing problems DownStream, or down the chain of collateral events?*

2. *Have you been looking at the DownStream issue (decreasing property values), or the deeper and actual cause of the problem (the Stream being blocked)?*

2

The Purge

"Take anything in your life that is Negative, and Physically rid yourself of it—Respectfully, with dignity, pride, and perhaps flowers. Toxins only work to poison your Frequencies, and your Frequencies are your link to change. This is step one in freeing yourself, and is the main application of Purging."

(I)

Purging and The Steps of a Successful Purge

Introduction to Purging

Purging[19] *is an involved process. In general, one should always work for balance in one's life, and should understand that when one is not balanced, then one will be vibrating at a Frequency that is not aligned with one's balanced self. Furthermore, this misalignment will not fix itself until one again finds balance, or until one works to create balance. A Purge is a dually-purposed process of both finding and creating balance, in which one will inevitably shift to a position from which one can then initiate Positive Change.*

[19] ***Purging*** *is the process of removing Toxins, and therefore Imbalances, from one's system, in order to create a position for one to initiate Positive Change.*

The Steps of a Successful Purge

There are four steps, or phases, that one will undergo when beginning and completing a Purge:

(A) Understanding What a Purge Requires

(B) The Tools of Purging

(C) Implementing a Successful Purge

(D) Finishing a Purge, and the Results

Finishing the Concept of The Stream Analogy

Before we continue, please note how it was intentionally stated that the effects of the Imbalance in the Stream were not immediate, but rather, over the course of time. This implies that multiple forces (the multiple energy Systems that the Stream is connected to) were acting upon the resulting state of the Stream.

(II)

(A) Understanding What a Purge Requires

Purging, and an Application of Chapter One

To truly Purge, one must first come to an understanding that "oneself" is in fact a creator of an infinite amount of Systems (within one's immediate connectedness) and that "oneself" has the choice of which Physical Systems that one chooses to partake in. There is an infinite Universe full of Systems surrounding "oneself", and as a Being of Consciousness, one can move freely between the Systems by choice and thought (unlike the hypothetical Stream). Beings of Consciousness also have the ability to understand and act upon any given System (learning and consciousness) so long as one has a Properly Maintained Railway Model (or connection to the Physical Realm of which the Systems exist). If one does not have a Properly Maintained Railway Model (an Imbalance), then one will not have a freely flowing Stream, but a blockage instead, which can lead to Depression, anxiety, worry, tension, and other such Negative

Imbalances, therefore negatively affecting ***The Natural Peace of Energy***[20]*.*

Toxic Entities and Purging

In a basic sense, the concept of a Purge is to rid oneself of every genre of Negativities: people who reject the belief that they do not need to help the System as a whole (including oneself), substances that are not Respected by the user (drugs, alcohol, etc.), counterproductive activities (overworking), disrespect (i.e. overeating, extreme selfishness), and any other Toxic entities.

[20] ***The Natural Peace of Energy*** *(or the Natural Flow of Energy) describes a state of balance in which energy is not hindered by blockages, and can freely flow through entities of the Physical Realm.*

Purging Negative People

In the case of Negative people, they will require a Purge of their own, and one is not abandoning them by purging them, but is simply instead making a choice to help oneself first, in order to then be in a position where one can help others. Step one is always to help oneself, but not in a selfish way — no, quite the opposite. You too deserve just as much Respect as do those around you, and therefore you too should be taking care of yourself, as should they. This is the duty we all have as humans, something that we all invariably have in common (The BPH). However, this is not a free pass to be selfish.

***NOTE: In this book, those that are Negative can hereafter be considered Toxins. They do not aim to help you in your successful growth and enlightenment, but instead aim to ruin it, whether they mean to or not!*

Food For Thought

As a general rule, Negative people lead to Negative results, and Positive people lead to Positive results.

Understanding and Identifying a Toxic Environment

If one is feeling depressed, trapped, or lost, one can safely assume that there is some source of Toxins holding one back from growing, or feeling at peace. One should then immediately put into works the concept of a Purge, and begin to rid oneself of the Toxins one is surrounded by. The first step in identifying a Toxic environment is fully understanding Toxins, and what Toxins mean to oneself. This ability to identify Toxins will be explained further in the next section of this chapter, under "Wants". For now, let us understand excuses, and how they can negatively impact Positive Change.

Excuses

Excuses[21] *are defined as any reason that one "is not able to" do something, which is inherently invalid since the Higher Realms are limitless, and the Physical Realm is simply a Manifestation of the Higher Realms. If one has just produced a separate reason for why one "cannot" Believe the aforementioned statement, then one should now recognize that one is engaged in a Negative thought pattern, and can begin to work to change one's Comfort Zone, or one's limitations, that one has fallen victim to. In practical terms, in our modern age, technology has a solution for any and every problem, in effect giving oneself infinite abilities.*

---

[21] ***Excuses*** *are any reason one cannot be able to do something. Excuses are always invalid, and a sign of being "stuck".*

To elucidate on the power of excuses, let us revisit the Comfort Zone[22]*, and let us define it as a hypothetical Box. "The Box", that one's Comfort Zone creates, will have walls made from statements such as "I can't because" and other such self-defeating phrases.*

The Box

The Box[23] *outlines the "walls", or limitations, of the mental structure one has set oneself within, and is strengthened by one's own thinking patterns. The Box is the biggest obstacle when it comes to Positive Change, and can certainly be overcome through Proper* **Discipline**[24]*, and a*

[22] *The Comfort Zone has been aforementioned, but is defined as the mental limitations oneself is currently self-defined by.*

[23] ***The Box*** *is a visual example of one being "trapped" in a Box by the "thoughts" that one has put up as obstacles to change.*

[24] ***Discipline*** *is the controlled practice of Focus.*

utilization of **Complete Focus**[25]*. In short, The Box is a figurative representation of mental obstacles.*

Excuses (Continued)

A commonly practiced societal excuse is "I am too old", yet there are examples of humans beyond middle age that are still accomplishing incredible athletic, mental, and "Box-breaking" feats. These Beings are not confined to The Box, and have challenged themselves to break free of limitations by stepping out of their Comfort Zone. Furthermore, any Being is capable of becoming aware of, and actualizing, these liberating concepts, so long as learning is practiced, clear **Wants** *are established, one Purges correctly (refining one's Frequencies for*

---

[25] **Complete Focus** *is a state of Discipline in which there are no Distractions from the entity one is choosing to believe at the time of which the Focus is being applied.*

the purpose of the Want), and Complete Focus is utilized.

Food For Thought

Creativity can solve any problem.

Understanding How to Become Limitless, In Order to Have a Successful Purge

Humans all begin a Tabula Rasa, and form conceptions of "oneself" through experiences and Classical Conditioning[26]*. As a result, humans are led to form The Box (an image of oneself). However, one is never restricted to one's Box, since The Box is a self-made*

[26] *Famous psychologist Ivan Pavlov and his exemplification of Classical Conditioning is a key concept in the matter of human learning and adaptation. *Research more on Ivan Pavlov and his concepts of Classical Conditioning to understand how thoughts are a trained response, and can be conditioned or unconditioned.*

conception, and conceptions can be formed and unformed (Classical Conditioning). With this understanding, one has the ability to overcome the concept of The Box, depriving it of its power, and thus returning oneself to a Tabula Rasa (the limitless childhood state that provides infinite room to grow, change, and develop).

Food For Thought

One can always create or destroy one's own Box.

The Impossible Has Been Proven Possible

The Wright Brothers were told they "could not" fly, yet they did; Lance Armstrong was told he "could never" race again, yet he did; humans thought they "could never" go to the moon, and the list goes on. The pattern is consistent: the impossible is possible, as long as a Being Believes it, and brings it to fruition. Once an impossible task has been proven possible, in the eyes of society, it becomes possible. Following this observation, one can then reverse engineer an understanding that anything is possible, so long as it is completed in the Physical Realm.

Irrational Limitations and Utilizing Complete Focus

The very diction "I cannot" is a Physical Manifestation of "you" thinking to yourself that you "cannot" do something, which then makes it conceivably impossible to do. Instead one should say, "I can try to", later refine this to simply, "I can", and finally evolve this to, "I am doing". This process will allow one to discover that one can do anything one wishes to Manifest, and that Words are a very powerful form of Manifestation, and should therefore be Respected and considered as such.

***NOTE: The only Being that can limit oneself is oneself, and any limitations one experiences will stem from thought patterns such as "I can't* *because", "I shouldn't because", and so on. These words (vibrations) are, by nature, limiting, and are considered Distractions, as they imply one is not utilizing Complete Focus towards one's Wants. Again, you are what you Focus on and words are a powerful form of energy.*

Food For Thought

In a scientific sense, a verbal word is the result of the vibration of one's vocal chords, making a noise that vibrates through air waves, as well as through one's Physical Being. Thus, they too have their own energy and impact in the Physical Realm.

Concluding "Understanding What a Purge Requires" and Intentional Forgetting

In order for Positive Change to occur, and to complete a successful Purge, one will practice "Intentional Forgetting". ***Intentional Forgetting*** *is the practice of forcing oneself to*

Focus on Positive entities (a practice that can be implemented after one has recognized a Negative thought pattern) and to, as a result, forget about all other entities (thoughts, past experiences, actions) that do not pertain to the Positive entities one is focusing on. This grants the awareness of one's current mental direction, the ability to refine and tailor one's thought patterns (or to Completely Focus them), and the absolute Manifestation of all things pertaining to one's Wants (a key practice in order to complete a successful Purge). Intentional Forgetting also ensures that one will no longer waste one's energy (since it entails eliminating "limiting" entities in one's thoughts) wherein that energy can then be redirected towards the goal of a successful Purge with Complete Focus.

(III)

(B) The Tools of Purging

Wants

One will usually get what one Wants, because what one Wants is a Focus of one's thoughts (which will eventually become Manifested in the Physical Realm). If one does not know what one Wants, then one will never get what one Wants, because one will always be Manifesting a Want of not knowing what one Wants! Again, this refines the previous statement, "You are what you eat", to a new understanding of, "You are what you Focus on"[27]. Wants allow one to define the guidelines of one's individual toxins, and answering the question, "What does one truly Want?" allows oneself to begin to define one's wants, and gives a clear direction towards the overall goal of creating a Natural Peace of Energy. For example, if one is confused about a topic, then one will most likely be confused by

[27] *On your own, research the term "Self-Fulfilling Prophecies" to further understand the concept of "you are what you Focus on".*

the outcome, and will never be able to Purge, because one will have no clearly defined Toxins or Imbalances, and will never understand what they need to Purge in order to free themselves!

Establishing Complete Focus

Establishing Complete Focus is essential in Maintaining one's Wants. The Wants of the Higher Self and the Focus of the Physical Self are always in a pas de deux: when the dance is practiced and perfected, the results are a smooth and effortless display between the Higher and Physical Selves (a Maintenance of the Natural Peace of Energy). However, when each entity is not Respected, we find ***Distractions*** *can enter the process, and slow if not halt a Purge entirely. This is why it is important to understand that the utilization of Complete Focus is essential in Purging. For emphasis, establishing Complete Focus involves ridding oneself of Distractions, and Distractions can be* *defined as anything that do not have to do with one's Wants.*

Ego

For the purposes of Positive Change, an ***Ego***[28] *is identified to be a non-Physical location from which a Want (not a need) of one's individual Physical Self stems from. As described in Maslow's Hierarchy of Needs, there are only 3 essential needs of the Physical form (food, water, and sleep). Everything else can be considered a Want, and can be identified to stem from one's Ego. If left unchecked, one's Ego will become a Distraction towards creating Positive Change in oneself, as well as on a massive*

---

[28] *Sigmund Freud expands upon this concept, as do many behavioral psychologists, but for simplicity, and for the sake of Positive Change, one should identify an Ego to be a non-Physical location from which a "Want" (not a Physical need) stems from.*

scale. After one balances one's Ego, one will then be able to "tune in" to the Egos (or Wants) of others. Honing the ability of checking one's Ego will allow one a vital skill in Creating Positive Change: the ability to see other Beings in their ***True Form***[29] *and their* ***Current Form***[30]*. With this ability, one can then be in a position to assist that Being with Positive Change.*

---

[29] *The* ***True Form*** *of other Beings is the form in which they Manifest when they are balanced.*

[30] ***Current Form*** *is the alternative form to True Form. It is the form one is currently Manifesting. Current Form is determined by the Being's knowledge, how that Being's Railway Model is functioning, as well as that Being's understanding of other concepts in this book. For example, perhaps the other Being is unaware of its Ego (leading to some psychological "disorders", i.e. Narcissism), or its own consciousness, and therefore that Being will Manifest in its Current – unbalanced – Form, as opposed to its True Form.*

When one is capable of seeing another Being in both their True Form and Current Form, then one may be deemed successful in checking one's own Ego, and will be able to then (without the Distraction of one's Ego), utilize Complete Focus on the target Being.

***NOTE: One should be careful not to suppress one's Ego completely (for the sake of Maintaining Self-Respect, and one's own Railway Model), but can practice doing it in increments, or in practical use. One's own Wants are important too, but beneath the umbrella of, everything in moderation!*

Controlling One's Emotions in Order to Contain the Ego

The inability to control one's **Emotions** *(or a Being's perception of a particular feeling) typically stems from either misunderstandings (of one's own System in the given situation, or of* *the connectedness of other Systems in the given situation), or of a Physical Imbalance (not enough food, water, or sleep). Therefore, in order to better control the way one responds to one's emotions for a given situation, one must have a clear understanding of one's True Form and Current Form, and a Proper Railway Model. Understanding one's True Form and Current Form grants one the ability of controlling one's Ego (or one's Wants), and requires a great deal of Discipline of one's thoughts. Understanding the two entities and the distinction between them will allow one to be able to Manifest one's True Form, regardless of one's Current Form. This can greatly help in interpersonal relationships, stressful situations, and long-term success.*

For example, Sally Sue tells Billy Bob in a conversation that she no longer wishes to engage in a relationship due to her life being so chaotic. Billy Bob could react by perceiving emotions of hatred and resentment, but if Billy Bob is balanced (experiencing his True Form), is not thinking from his Wants (or his Ego, a characteristic of the Current Form), and is able to understand Sally Sue is not balanced (therefore not experiencing her True Form), he is able to change what he Manifests, and instead tune into Sally Sue to realize that she is under a lot of stress and is currently Imbalanced. He can then take it a step further to fully remove himself from the equation of this situation (by removing his Ego), thus granting him the ability to look at Sally Sue without any Distractions, and to see what her Immediate needs and Wants are. This is a practice of Respect, and will allow him to handle the outcome of the situation in a much smoother way— since he is able to understand where Sally Sue is currently at (Imbalanced, experiencing her Current Form instead of her True Form, and therefore in need of her own Purge).

Complete Focus Continued

In order to create Positive Change, one must always Want to create Positive Change. This will in turn create a self-sustaining Stream from which one's Physical needs will be eternally provided for, since any Want that one has will always require a preliminary step of Maintaining oneself. Again, one must first help oneself before one can be in a position to help others.

With this System, one will always be aware of one's needs, and the fact that those needs will need to be fulfilled at some point in the process of pursuing one's Wants. After the Physical needs are fulfilled and self-sustainable, Complete Focus will follow suit, as will also the ability to Properly utilize something called "Gauges".

Gauges, and a Brief Introduction to Remaining "In the Moment"

Gauges[31] *can be compared to the speed gauges in a car: the speed gauge tells one how fast one is currently traveling at that particular moment— not how fast one was going, or how fast one could go. Therefore, Gauges are exactly that: a snapshot of a current moment, in that moment's purest form. Gauges are an effective tool in understanding how to be "in the moment", or present, and are a practice of Complete Focus.*

When one utilizes Gauges, one simply looks at a situation for exactly what it is, without influence from the Ego or Distractions, and with the

[31] ***Gauges*** *are the truest and purest form of understanding a current situation, without the hindrance of the past, the future, the Ego, or any other Distractions.*

understanding that one can only control the forward result of the proceeding moments, and not the past result of what has already happened. Gauges involve a constant assimilation with oneself to the surrounding Frequencies, and an openness to go with the flow, accept change, and appreciate what it is that one has been presented with, in real time. One can always utilize Gauges for Positive Change, in any given situation, at any moment, for there is always Beauty to be found, whether one presently realizes it or not.

The Distractions of the Past and the Future

When one does not Properly Maintain one's Physical Self (an improper Railway Model), then one will never have a full connection to the Physical Realm, and therefore will never be able to be fully "present" (literally, in the present) which will in turn lead to a cascade of Negative events. For example, ***Worry***[32]*, a concept that typically involves a Being projecting one's mind to a future or past of possibilities that aren't presently happening, entails one is not Focused on the here and now (the Gauges – that which the camera is taking a picture of).*

Worry is one of a few powerful Distractions and signs of Imbalance, and is a common theme in athletic development. In athletics, if one is performing an action, but "worries" about a past attempt (in which the activity was deemed to be incorrectly performed), then one will most

---

[32] ***Worry*** *(a varying degree of Anxiety) also holds roots in Buddhism and Neuroscience. In Buddhist belief, Worry can be considered a state of Imbalance in which energy is disrupted in its flow between the Chakras (the energy systems of oneself). Similarly, in Neuroscience, Worry is proven to release a certain hormone that blocks Proper functionality of one's body. In both beliefs, the result is an inability to function in a Natural State of Peace.*

likely Manifest that worry (a self-fulfilling prophecy) and in turn misinterpret details (misread the Gauges) of the present action. This Focus on the past is considered a Distraction, as it does take one away from one's Wants, which, in a balanced System and in this hypothetical situation, should be directed solely upon performing the action correctly (through Complete Focus). Conversely, if one is always envisioning the future and focusing on what one is trying, or hopes to do, as opposed to what one should be doing, then one is misinterpreting one's Gauges, instead of actively reading them This is due to not being fully present, and thus not utilizing Complete Focus, wherein one will never be able to bring that particular envisioned future to the present (discussed in the next chapter, "Recreating Oneself, and Quantifying Results").

***NOTE: It is important to understand that Past and Future are not always Distractions, as they need be utilized Properly in order for a successful Purge. Utilizing knowledge of Past and Future can be understood to fall under the umbrella statement: there is a time and a place for everything.*

Complete Focus Finalized

Complete Focus is a state of existence without Distractions. That which is not ***Priority***[33] *(Distractions) must become irrelevant: if it doesn't help you read your Gauges, then it is considered a Distraction.*

[33] ***Priority*** *is a logical listing of what is important, or what one Wants the most, and, in the occasion that there is more than one Want, is arranged in order of importance to that individual Being.*

Ridding Oneself of Distractions

The process of ridding oneself of Distractions may cause internal conflict between one's Physical and Higher Self. This is because physical attachments can sometimes be Negative influences, and one's Physical Being may sometimes conflict with the Higher Self's ***Logic***[34]*. Emotions felt by the Physical Self's* ***Heart*** *(a term that denotes the Physical entity of one's emotional feelings, which is experienced by the Higher Self), can sometimes be misaligned by Negative Influences. After all, one is considered Imbalanced while Purging (and therefore implied to be surrounded by Toxins), therefore when the Heart is misaligned it may be improperly perceiving emotions. This can*

[34] ***Logic*** *consists of calculations made in response to the Physical Realm that are concurrent with one's priorities.*

conflict with how the Higher Self understands the actions it needs to relay to the Physical-Self.

For example, there have been known occasions where an individual "Wants what is worst for them". This is an exemplification of one such occurrence where the individual is Toxic, and needs to Purge, regardless of what Negative interactions the heart wants. As one initiates a Purge and begins to rid oneself of these Toxins, a balancing will begin, and internal conflicts will subside. Until then, however, the conflicts between the Physical and Higher Self will lead one to cognitive dissonance (as explained by psychology), and will be understood to be the expected result of an Imbalance.

Since it is implied one is Imbalanced throughout the Purge, it is also understood that one cannot yet fully utilize Complete Focus for extended durations until after a successful Purge is completed, meaning the process itself of detoxifying these entities may take time, and will definitely take Patience. This should not discourage one in attempting to Purge, but more so, should become an inspiration, since a new level of emotional awareness is the end result of a Purge. This is the reward for the effort and Work that goes into a Purge, especially because sometimes it may require a seemingly endless amount of time before balance is restored. With these things in mind, one should mentally prepare oneself prior to a Purge with the understanding that a Purge is a time of great difficulty, Work, and effort, but is only temporary.

Defining One's Individual Wants

When establishing one's Wants, one should take much time and deliberation to do so, with an understanding that there can be NO excuses that limit the chosen Wants. From this point on, excuses do not exist, and are not an option. Furthermore, one's Wants are a compilation of one's own thoughts, so when beginning to establish one's Wants, one should refrain from letting others know, or even allowing others to influence one's decisions in this matter. They are called "your" Wants, not "others'" Wants. One can trust no one until after a Purge is completed, and should therefore keep to oneself when establishing one's Wants[35]; not in a Negative way, but more so in a way where the process is optimized and most efficient. This does not mean one should stop daily activities, but instead means setting physical time aside to

[35] *If one is at a higher state of Positivity when initiating a Purge, one may have a Positive influence (or influences) already established, but these can be Toxins in disguise, holding one back from one's "True Form". One should test an influence (after one has identified one's own Wants without external influence) by discussing these Wants with the believed-to-be Positive influence(s), and seeing if they are in support. This process of "testing" is in line with the methods of identifying Toxins, and is an effective method when utilized correctly.*

meditate (a form of uninfluenced thought) on some of the aspects of a Purge. A good starting question is, "What do I Want to do with my life next?". One has certainly already been doing things with one's life, but if one is looking for Positive Change, then one understands it is time to move on to the next chapter and begin to define exactly what one Wants that chapter to be. The process of establishing one's Wants is a practice of Complete Focus. Now begin to ask yourself: Do you Want to become an astronaut? Do you Want to write a book? Do you Want to move to Hawaii? Do you Want to lose weight? Do you Want to move past a relationship? Do you Want to chase a dream? Etc. There is always an endless amount of possibilities (as was stated earlier in proving the impossible is possible), and one should invite oneself to be daring in this process of self-discovery. Pick a Want that is intentionally not easy, and allow oneself to fail hundreds of times before creating the precise Want one had originally sought! Failure is an incredible teacher, and is inevitable, even necessary, in the process of defining one's Wants.

***NOTE: Defining one's Wants is not an overnight process, and should NEVER be restricted by a time frame. This is why when actualizing the creation of Positive Change, one must challenge oneself to truly step out of one's Comfort Zone in order to refine what it means to be "oneself", or to even change that conception entirely.*

Defining One's Individual Values

*Once one has established Wants, one can begin to define what **Values**[36] one needs to possess in*

[36] ***Values** are the particular skillsets that are required for Manifesting a Want, as well as the substance (qualities, vision, etc.) that makes up the Want one is establishing.*

order to Manifest those particular Wants into the Physical Realm. For example, one may desire to be the best possible fireman[37] *the world has ever seen, but is starting at a point in which one knows nothing about being a fireman. This place of absolute uncertainty is the greatest place to start, and will be outlined further in Chapter 3.*

Food For Thought

One should not fully understand what it takes to become the best possible fireman the world has ever seen when first setting out for that particular want, and additionally, one, by implementing a Purge, has already taken steps towards accomplishing this Want. It is all about the journey, not the destination!

[37] *Herein, we will use the hypothetically-defined Want of "becoming a world-renowned fireman". You are free to put your own Want into this placeholder and apply the same practices.*

Being Clear With One's Wants, In ALL Moments: A Practice of Complete Focus

Being clear with one's Wants in all moments is key to the process of the Purge. It entails one is Maintaining a Proper Railway Model. It is key, because one needs to remain consistent in order for true change to set. If one does not remain consistent, one can easily fall back into one's old, Negative cycle, and a new Purge will need to be initiated. Consistency is another tool in the Purging process, and requires the development of methods that Maintain this consistency, such as frequently reminding oneself of one's individually defined goals.

Finding Balance During a Purge

If one has never before found balance, then one should take extreme measures in their Purge. Separate oneself completely from all influences, and any resemblance of the structure defined by one's individual "Box". This will allow for a completely pure Purging process, and therefore a complete cleanse, and a complete chance of success.

(IV)

(C) Implementing a Successful Purge

Recap of Concepts So Far, and Initiating a Purge

Since thoughts originate in a Realm other than the Physical Realm, they are considered to stem from the Higher Realms. Thoughts are not Physical; they are not tangible, nor do they have mass— but they do have an effect on the Physical Realm. This leads one to speculate about the truth behind one's thoughts, how the Physical Manifestations of oneself come from one's Higher Self, how one guides one's Physical Being with one's thoughts, and, the most important realization, how one's thoughts originate at one's Being.

Toxins are Negative, and therefore oppose that which is Positive. To initiate a Purge, one must first understand what is considered a Toxin in one's immediate connectedness, as well as immediately Work to Purge these Toxins. Being surrounded by too many Toxins ruins one's ability for a Properly functioning Railway Model. Therefore, ridding oneself of Toxins effectively allows for the increase in one's ability to channel Positive Energy.

Food For Thought

Negativity has a "weighing down" effect, and Positivity has a "light" effect. One may feel light when experiencing Positivity, and heavy when experiencing Negativity.

One Will Manifest Changes to the Physical Realm Through Beliefs

Believing something isn't magic. Believing something is a concrete matter of choice, strengthened by thought patterns, and changed through Discipline. One must consciously choose to Focus on (and also Believe) an entity in order for it to become Manifested in the Physical Realm, therefore one must consciously choose to Believe one is engaged in a Purge, and choose to Believe that it is working in a consistent manner for the best results. One can, and should go to extreme lengths in their Purge by Physically expressing signs of passion, open excitement, and a welcoming for the new life they will have afterwards. This will help to improve the efficacy of the process.

Express Oneself!

Tell others you are Purging! Explain to them the process! Those that are Negative, tell them you think they are Negative and why you Believe that! Ultimately, it is not about their response, but more so that you are speaking your Beliefs into the Physical Realm, and Manifesting the concepts of a successful Purge, but this does not grant one the ability to go around making enemies. Purge with a gentle care for others' feelings. Remember, consistency amongst all Systems in one's immediate connectedness is key, and requires Positivity to be a crushing force to external Negative influences! This way of self-expression will expedite the process of the Purge, and is quantifiable by The Results Equation[38]*.*

A Moment to Celebrate

Allow oneself to take a moment and recognize that one has just solidified a major (if not already previously known) discovery: that one is the creator of one's own manifestations, and therefore one's own life. More so, one simply need use the tools that are innately at one's disposal in order to manifest Positive Changes! As was previously stated: "Oneself is in fact the

[38] *The Results Equation will be discussed in Chapter 3, but for now understand that it defines a central dogma: the factors one puts in to a particular system will determine the result. In this case, the more energy directed towards expression is a factor.*

creator of an infinite amount of Systems (within one's immediate connectedness) and oneself has a choice of which Physical Systems one chooses to partake in", meaning one's own life is completely in one's own control. This alone is an infinitely amazing ability: the ability to create new Systems, or to create change. This ability is also essential in beginning to take steps into a new life that one will create for oneself, with the help of new Positive influences found during or after a successful Purge.

Time to Purge

Here is the challenging part. Here is the part where one must step out of The Box. Here is the part where one will not feel a "smooth" transition, or sunshine and rainbows. Here is the part where one must utilize consistency. Here is the part where one learns of true Toxins. Here is the part where one practices Complete Focus. Even with all the tools, a Purge was never said to be a comfortable thing. Thankfully, the result certainly is. Allow yourself ample time when initiating a Purge, and take small baby steps. Rome was not built in a day, and going in with the expectation that it will take a while increases your chances for success.

***NOTE: Good luck in this endeavor, and do not be fooled: if you have to question if it is truly over yet, then you must understand that you are far from the end.*

Food For Thought

Knowledge is Power.

And now, we Purge…

(V)

(D) Finishing a Purge, and the Results

Re-Introduction to Complete Focus

It is formerly understood that thoughts are another form of energy, and energy (and its resulting vibrations) is measured in Frequencies. With the concepts of Belief and Focus, one can now evolve the concept of Manifestation to say that the Higher Self may in fact create Frequencies in the Physical Realm through various modalities (such as "thoughts"). This Manifestation occurs by Focusing one's entirety (one's thoughts) on a particular entity, and this process is known as Complete Focus. Complete Focus is the absolute key in Manifesting Positive Change. It is the final tool one needs in order to complete a Purge. It entails one still allows Positive influences, but ignores (Intentionally Forgets) Negative influences.[39]

---

[39] *It was previously stated that one should assume one knows nothing about Positive and Negative influences until*

Characteristics of The End of a Purge

One must always understand that when one is experiencing a change in one's life, one will experience a period of perceived difficulty. Change implies a new "flow" of Frequencies, which involves physical and mental adaptations. Physically one's body will vibrate at a new energy, and will therefore feel different, and Mentally one's Higher Self will think differently, in effect changing one's Physical Being. Because we can be Beings of sentiment, the end tends to be so difficult, and we may need to come to peace with the things and the actions

---

the end of a Purge. Now that one has finished a Purge, one will have learned how to distinguish between Positive and Negative influences, but should keep in mind one will fail at understanding this difference many times before one finally is able to produce a refined distinction between the two. One should engage in trial and error, and utilize research for the sake of expedited learning.

that have ensued the Purge (i.e. letting a friend go, a lover, or having a difficult interaction). To come to peace with these things, one need only to accept change and to allow oneself to evolve, and it is through Respect that these changes remain in power.

Time

***Time** is a reference of change in the Physical Realm, which is usually measured in increments. If something hasn't changed, and an increment of time has passed, then it can be assumed that a force has been working to preserve the System. Usually a Being does this, inputting energy into its preservation. This reference of change is true for both Positive and Negative cycles, and is why it is important for one to recognize when one is in a Negative cycle, and to subsequently initiate a Purge.*

Food For Thought

One could be the very force preserving the Negative cycle!

Following a Purge

A successful Purge will entail that one has now become balanced, has stopped a Toxic System (a Negative cycle), has experienced much change, uncertainty, and inner tension, no longer has any Negative influences in their immediate Railway Model, and is now at a place where one can begin to develop and create an entirely new System, with entirely new Frequencies, without the distraction of returning to the old Frequencies (or the old Comfort Zone).

Additionally, one should find comfort in this understanding: although this Purge may have required one to go through the potentially uncomfortable process of removing Negative people from one's life, one would never would have had a chance to find new Frequencies (those that will fill one's Being with Positivity) if one hadn't originally Purged the Toxins that were blocking one's Railway Model.

Purge Completed

Now that one has ended a Negative cycle, one can recognize that one is at a starting point (known as ***Zero***[40]*), and that Zero is a place of infinite possibilities.*

---

[40] ***Zero*** *is defined as the representation of the point in time from which one can begin to create a new cycle.*

Welcome to Zero.

3

Recreating Oneself and Quantifying Results

"I stopped looking [for Positivity] and let it find me. As for [Negativity], it's part of life, but has no place in my life. Love all, and let those that hate, keep hating. Decide from this moment what kind of person you want to be, and live accordingly." – Unknown

(I)

"The After Purge"

Introduction to "The After Purge"

Welcome to Zero, the place where infinitely everything is possible, and any steps in any direction take you down new paths to your beautiful new life. With the Purge successfully completed, one can and should now take a moment to sit back and relax. The difficult part is over! Time to celebrate! Yes, celebrate! Not with any one (unless you want them there—Positive people only!), but with you, yourself, and you! You are the victor of this great challenge, and you have now taken the first step in taking back your life! Hug yourself, assure yourself that it is okay, and let yourself know that you are now on to the next step in this wonderful journey of happiness. There are only better things to come...

Things to Note in "The After Purge"

After a Purge, one will undeniably feel differently than one has ever felt before, and be in a completely new place (known as Zero), with no chance of returning— Yes, "no chance of returning", because the idea of "going back in time" (say we decided to text Sally Sue again to rekindle things, or Negative Nate to see how they were doing) would be ineffably difficult. There may be a redefined relationship with Sally Sue or Negative Nate, but never will things be the same as they were prior to the Purge (due to a few reasons). On the one hand, one is now a completely redefined oneself: one may find one no longer holds interest in the activities that had previously bonded one to those Negative influences. On the other hand, since a moment in Time is defined by the Frequencies that were occurring at the point during which it was measured, it would be ineffably difficult to recreate every single possible combination of

Frequencies that went in to that moment. Notice how it is not said to be impossible? It is important to leave open all possibilities with the understanding that we, as a human race, do not know enough in order to yet deem something "impossible". That would be very ignorant, inane, and foolish, especially after only a meager 5,000 years of recorded history (starting with the Sumerian Cuneiform Script)! However, for the sake of Positive Change, and to avoid relapse, we must create the Belief that it is not possible to go backward in Time due to its ineffable difficulty, and instead only forward in this journey of Positive Change.

***NOTE: Releasing the weight of the past will allow one the freedom of the future...*

Extended Durations of Toxicity and the Rehabilitation Period

If one has experienced an extended duration of Toxicity, then it must be understood that one will need a ***Rehabilitation Period***[41]*. The Rehabilitation Period allows oneself time to return to center naturally, through the conduit of Work (a measure of Time and Focus), and the success of the Rehabilitation Period will be defined by The Results Equation. The Rehabilitation Period has no specific guidelines. It simply should be a time of processing, slow growth, and settling into new habits.*

---

41 *The* ***Rehabilitation Period*** *entails an allotted amount of time in which one allows oneself (through Work) to return to Zero.*

The Result of a Purge, and Shifting Focus to Recreating Oneself

As a result of Purging, one will come to the understanding that there is a connection between the Physical, and Higher Realms. One will begin to understand one's thoughts (energy from the Higher Realms), how these thoughts enter the Physical Realm (through one's Railway Model), the role of The Ego and Distractions in one's life, and how one's thoughts are expressed through the energy that is surrounding "oneself" (one's immediate connectedness and Railway Model). The efficiency with which one's thoughts Manifest from the Higher Self into the Physical Realm is entirely dependent on the state of one's Railway Model, and how Properly it is Maintained (as aforementioned).

The After Purge is where one now (with a fundamental understanding of one's Railway

Model) can further explore one's own capabilities, and can become the creator of one's own physical destiny, with a set of Wants, no Negative influences, and the tools to do so.

(II)

Recreating Oneself

Just do it, and DO NOT give yourself time to think Negatively. There can be no Distractions.

**Some of the concepts in this section are not discussed in depth, for sake of brevity.*

Introduction to Recreating Oneself

Systems are cyclic, and therefore, since one had been living a certain way with certain Frequencies for as long as one had been doing it for, one now needs to recognize exactly what is required to balance out that old Negative cycle (The Rehabilitation Period). This will allow one to not only initiate a Positive Change, but to also successfully finish creating one. Stay Persistent and Vigilant in this process!

Since one is now well underway in becoming "that world-renowned fireman", it is at this point that one needs to understand the importance of building a strong foundation. A strong foundation is the outcome of layers and layers of Positive results that have been carefully implemented with Patience, over the course of time. The strength of these layers will be in direct correlation to The Results Equation (discussed in-depth later in this chapter).

Breaking Things Down, to Build Them Back Up: Start With a Strong Foundation

Now that one has Purged, is in a state of new beginning (Zero), and has ultimately ended a Negative cycle, one is readily in a position to create a new cycle! For emphasis, this new cycle will be one that is tailor-made for one's Wants, and the success of this new cycle will be the result of a Proper practice of Patience, as well as the building of a strong foundation.

The Results Equation

The ***Result(s)***[42] *of a given project (i.e. a Purge or a Creation of Positive Change) is directly proportional to three factors: Focus, Energy, and Time.*

---

[42] ***Result(s)*** *are the outcome(s) of a system at the time at which it is measured. A* ***Product*** *is a Physical, tangible result of a system.*

Before defining these factors in the context of The Results Equation, it is important to understand that The Results Equation is a conceptual tool, with a vast array of practical uses, not outlined in this literary work. In summation, The Results Equation outlines that every System will have a different amount of Factor inputs, which will proportionally affect the Quality and Quantity of the Results (outputs).

Food For Thought

When applying these concepts, and in all occasions, ensure to balance quality over quantity.

The Factors and The Uses of The Results Equations

The Results Equation *reads: R = (F [+ or -] Q) x (E [+ or -] Q) x (T [+ or -] Q), where R is Results, F is Focus, E is Energy, T is Time, and Q is the Quality of those variables (which* *ranges from positive and negative quality inputs).*

The Results Equation Continued

The Results equation does a number of things, perhaps most importantly being that it provides a conceptual standard of determining what happens to the Result of a given project when Complete Focus is not utilized, as well as the effect of Positive and Negative inputs of Energy on the System over the given increments of Time. Again, this is not discussed in depth in this literary work, but understand that when one Wants to Manifest something, such as a Concept (a completed thought stemming from the Higher Realms) or an entity (a System in the Physical Realm), this equation always holds true. The extraneous variable of Quality plays a key role in the overall result, and it is why it is implied one should utilize Complete Focus in order to maximize Results. The practice of Complete

Focus entails there are NO Distractions and only Positive Energy inputs, implying a maximum amount of constructive energy. A maximum amount of Positive inputs can optimize the efficiency of the project, translating it into optimal Time consumption, and ensuring the greatest possible Quality of Result.

Food For Thought

Since The Results Equation always holds true, one can simply take one's Focus away from an old Negative System in order for it to dwindle and collapse. Remove the input, and the System will cease to perpetuate.

It is also important to acknowledge that in this removal, as the old System declines, it will Maintain its strongest roots the longest.

Maintenance of a System

Maintenance *is defined as the combination of Complete Focus, Patience, Work, and Respect* *that is put into a System, and will require less amounts of* ***Work***[43] *than the initial creation, so long as a strong foundation was implemented and grown. In order to practice Maintenance, one need not put in enough Work to achieve results, but one does need to put in enough Work to prevent* ***Entropy***[44].

---

[43] ***Work****, in the context of this book, combines Focus and Energy. Thus Work = Focus x Energy.*

[44] *The Universe has a tendency towards disorder. In the occasion of Positive Change,* ***Entropy*** *defines the principle that observes that a given system tends to disorder, and in our application, will collapse if no Proper Maintenance or Work is performed upon it. According to the laws of thermodynamics, energy cannot be created nor destroyed, only transformed (The Law of Conservation of Energy) thus there is in effect an infinite amount of energy that the Universe has, allowing for a system that can exist infinitely in a perfect harmonious balance (The Natural Peace of Energy), so long as The Results Equation is in check, and Proper Maintenance is performed.*

Understanding Maintenance will allow one to quantify a results-based, or entropic-based outcome of a System (so long as the inputs are understood).

Elysium

Once one creates a System in which the Natural Peace of Energy is unhindered, then one will in effect be in a place of Beauty and pure Positivity. This is known as ***Elysium****. Elysium is a System in which Positive Energy is so foundationally strong (and properly Maintained), that no Negative Energy can influence it. It is a System of Created Beauty and pure Positive input. This can be achieved following a Purge.*

How to Become a Diamond

Diamonds *are an immensely strong Physical material, and are the end result of a lengthy passage of time, involving immense amounts of* *pressure. In fact, diamonds (which are incorporated into tunneling drills due to their intense resilience) actually begin as something completely vapid of all form, such as carbon dioxide. Carbon dioxide, once subjected to intense amounts of pressure (pressure being an incredible force) over a long enough duration of time, results in an immensely stable "crystal" structure, which is to highlight the ability nature has of creating beauty from a gas (carbon dioxide).*

The same is true for oneself! To become a Diamond, one must apply an extreme amount of pressure in order to strengthen the layers of bonds that hold oneself together. This pressure can stem from the challenges that one puts oneself through, and the limits that one pushes oneself to. First, one must understand one's "bonds", break them (if they are weak), then rebuild them in order to strengthen one's foundation. This is truly the process of recreating oneself, and is sometimes repeated many times until one has created the exact result they one was looking for. It involves deliberate trial and error, and an openness to the greatest teacher of all: ***Failure***[45]*.*

---

[45] *With the correct Positive mindset,* ***Failure*** *is by far the greatest achievement of all. Failure implies one has attempted something, has not succeeded yet, and now understands something new about the venture they are trying, which in turn takes one a step further in the process of actualizing it. The only true failure is to repeat the same mistakes (as T.S. Eliot says, "Hell is a place where nothing connects to nothing"), or to never try at all.*

Food For Thought

If one were to ask a scientist a simple explanation of why a diamond is so strong, they would respond: due to the way the atoms are bonded. The strength of each individual bond is in a perfectly harmonious and patiently formed balance. This is why diamonds are highly symbolic for strength, transcending to the culture of marriage ('strength in bondage') in current American society.

In order to find balance, one must often "tip the scale" to test the limits…

(III)

Setting the System

Doing all the right things

Finding Balance, and a Conclusion of Concepts

In setting and Maintaining a Proper System, one must always be in a state of balance. One may "tip the scale" if one is curious to see the limits of balance, meaning they can add pressure to see how it affects one's balance, but understand that this may have adverse effects. "Tipping the scale" implies one may be subjected to the great teacher of Failure. Additionally, one should understand that Elysium is achievable, as long as one has an understanding of balance of Work, Time, Respect, Patience, and Maintenance.

Creating Positive Change

When one is creating a new Positive Change, one must always be in the act of visualizing what one is trying to achieve. The vision is everything, and Manifestation of the vision is the product of consistency. The vision is that which the Higher

Self creates, and then begins to Manifest in the Physical Realm through thoughts and the Proper alignment of one's Railway Model and one's Body. This in turn permits one to Properly channel the Frequencies of the vision one is trying to Manifest.

***NOTE: Since Frequencies are synonymous with the energy of the Physical Realm, does it not make sense that if one attunes oneself to the correct flow of Frequencies, or to the correct vibrations that align with one's vision, then one will eventually become the exact Frequency of that which they are working towards achieving? Yes! And by definition, change is different, therefore so too will one be different after a successful change has occurred. New you, new experiences, new Frequencies.*

Lastly, it is important to note that Positive Change is not about planning: planning always has a chance of improper execution. Rather,

Positive Change involves steadfast "Intuition" (explained below). Once one has correctly Purged the Toxins from one's life, then one will be capable of utilizing a Proper flow of energy (Natural Peace of Energy), and therefore a Proper use of one's Intuition and Gauges.

Intuition

Intuition *is one's ability to freely Manifest one's thoughts (or energies) into the Physical Realm without hesitation, tension, or any setbacks. Intuition is a sign of balance, and implies one understands what one knows, and what one doesn't know (utilization of one's Gauges) at any given moment, in order to attain Proper and immediate decision making-to-action.*

The Four Choices of Any System

Once a Properly flowing System is established, it need only be Maintained in order to keep it at its current level of functionality. If one wishes to grow it further, one may do so, but this is optional, since only a baseline Maintenance allows for the continuation of a developed System and the prevention of Entropy (in the hypothetical instance that no other forces are acting to destroy the System). This leads one to the realization that, once one is aware of a System, there are four choices one can make: one can Maintain a System, grow a System, destroy a System, or remove awareness from it, with Intentional Forgetting. In the case of Positive Change, one never considers destroying a System, not even in the event of a Purge, unless it is a System one has developed completely by oneself, in which case one then has the right to destroy it (as it has implied ownership), or if the System is producing a destructive or negative Result. This is a fine line, however, and it is safer to never consider destroying a System and to instead practice option four, intentional forgetting, in which one will instead allow it to dwindle, thus giving other Beings the chance to Maintain it, if they care about the System, and ensuring that one is not acting selfishly.

4

Positive Change on a Massive Scale

In this chapter, we will stray far from the traditional yin-and-yang (+ and -) concept of Energy, to speak more of its Manifestation through human nature. Step into an amazing vision of a world where Positive Energy thrives! Mutual Respect amongst humankind, and a bond through our shared quest for knowledge of our existence (for example, why have human beings been made aware of ourselves- 'the human experience'-, and other creatures seemingly not?), and individual growth through community, discovery, individuality and understanding.

(I)

The Positive Movement

Before we can truly set our sight on the art of growth, we must complete a "Positive overhaul" (a complete healing of all Negative energies in humanity), in order to create a freely-flowing energy circuit (in the Physical Realm) with a set Focus on discovery. After all, we (Humankind—all bonded by the Basic Principle of Humanity) are only as strong as our weakest link, and even our weakest links thrive in happiness.

The Positive Movement

The Positive Movement is a formal coterie that any Positive Being intrinsically belongs to. It is a call to action and a way of uniting those that are creating Positive Change in the world, in any way, small or large, noticed or unnoticed. There is a great pride amongst members of this movement, and "The Law of Attraction" will soon bring us all together.

The Law of Attraction

The Law of Attraction[46] *is simple: Positive Frequencies lead to more Positive Frequencies, and Negative Frequencies lead to more Negative Frequencies. Once one is vibrating at a purely Positive Frequency, one will in turn begin to notice many Positive Changes in one's quotidian*

---

[46] ***The Law of Attraction*** *is the Dogma that Positivity attracts Positivity, and Negativity attracts Negativity.*

life. Furthermore, once one has Purged, set their System, and is enjoying the unfolding, one will find that things become smooth and effortless, and that ***Blessings***[47] *seem to coincidentally happen!*

***NOTE: Since the energy one is creating with Positive thoughts will Manifest in very Positive ways in the Physical Realm, one will naturally "attract" new Positive opportunities to oneself, due to the Properties of the Physical Realm, Elysium, The Law of Attraction, and a smoothly flowing Railway Model.*

---

[47] ***Blessings*** *are defined as unforeseen bits of good-luck or Positivity, in any form.*

(II)

Positive Change on a Massive Scale, and Becoming a Positive Healer

"The planet does not need more successful people. The planet desperately needs more peacemakers, healers, restorers, storytellers, and lovers of all kinds." – Dalai Lama

Energy Work

Energy Work[48] *is the utilization of different vibrations in order to influence Physical structures. Tools such as singing bowls are often used for this, or singing in general*[49] *(again one's voice is simply the result of vibrating the muscles known as the vocal chords). There are different Frequencies at which Energy Work operates (which can be channeled depending on the purpose of the Energy Work), and therefore there are varying intensities (varying Frequencies) at which energy can and does exist. In our modern age, there are some doctors*

---

[48] ***Energy Work*** *comes in many forms: Meditation, Electric Stimulation Therapy, Reiki, Tai Chi, Massaging, Technology, etc.*

[49] *This is why music is so influentially powerful. Beware the vibrations one is exposed to in one's quotidian life (i.e. on the radio). They could be subconscious Negative influences!*

who are now experimenting with sound therapy, and finding incredible results.

An Energist

A skilled ***Energist***[50] *is a Healer, and operates in their True Form, free from their Ego. An Energist, being an Empath, also understands that Energy Work can realign Proper energy channels (fixing Imbalances) for a given Railway Model, and therefore has the ability to assist in healing the Railway Models of others.*

Using Frequencies In Order to Influence the Physical Self

Since the Physical Self is bound by the laws of the Physical Realm, it must be allowed Proper treatment in order to function Properly. The

[50] *An* ***Energist*** *is a Being that understands the flow of energy through a system, and how to guide energy for Positive healing purposes.*

Body can resonate at many Frequencies, all of which are important for different purposes, and all of which can be utilized at some time or another. Rest, for example, is allowing one's Physical Self to resonate at a lower Frequency (a Frequency associated with healing or restoring) – which is why at rest one's "Systems" are replenished (like recharging a battery). Scientists study rest by identifying the electrical waves the brain shifts between during different activities (such as when one is active, in comparison to when one is sleeping) and have termed five major states of brain activity: alpha, beta, theta, delta, and gamma. These states support the use of Frequencies to define emotions, since varying levels of electromagnetic charges trigger certain neurotransmitters responsible for various feelings. Frequencies and their ties to emotion, as well as the five brain states, are not discussed in depth in this literary work.

With this understanding (that the brain operates with Frequencies), one can become self-sufficient in the healing of oneself and the Maintenance of one's Physical needs (according to Maslow's Hierarchy of Needs) by exposing oneself to the Frequencies that are in tune with the result one is trying to achieve.

The Application of Positive Frequencies in Creating Positive Change

When creating Positive Change, Positive Energy needs to exist solely as a crushing force to Negativity. This does not mean that one should suddenly become arrogant, and think themselves a greater Being than those that are Negative, but more so recognize the state in which a Being exists, and know that the Positive Beings are in fact serving a higher purpose, and eventually will be helping the Negative Beings, whether it be immediately, or through the Positive

Frequencies one is creating into the Universe (and the ripple effect).

(III)

Further Necessary Tools for Positive Change

Introduction to Changing the Flow of a Moment

Once one has become practiced enough at creating Positive Change for oneself, one will also be able to do so for others by doing something called changing the flow of a moment. This is true for occasions in which one is very balanced, and in which one has a clear understanding of the subject for which one is targeting (an understanding of their Current Form and True Forms). For example, and in a general sense, when speaking to a Negative subject, one will experience tension in the Frequencies of the Negative person, and therefore an Imbalance in the exchange between oneself and the Negative person. This underlines a huge miscommunication error that is often overlooked when two Beings are trying to interact. When the Frequencies don't line up, or an understanding of one another is not shared, then there is no true communication, which can result in a Negative cycle (depending on that particular Being's understanding of the content of this book – there are occasions where misunderstandings are resolved, and other occasions where they are not, leading to many Negative thought patterns (judgment) and Energies).

Encountering a Deeply Negative Being

When a Positive Being encounters a truly Negative Being, a sort of combat between one's own Frequencies (the Positive ones) and the Frequencies of the person with whom one is interacting (the Negative ones) begins. The result of this combat is entirely in one's control, and is dependent on one's proficiency in initiating the ***A, B, C, D Methods of Positive Change.***

The A, B, C, D Methods of Creating Positive Change: When Dealing With Negative Entities

In the aforementioned situation, in which a Negative and a Positive Being are engaged, the Positive Being is able to explicate four primary courses of action: A) Attempt to influence the subject Positively through Facilitated Learning, B) Allow the subject to express themselves, and do nothing at all, C) Purge, or D) "Tune in" to the Frequencies at which the subject is operating.

In the occasion of creating Positive Change, one would always attempt A first, but if the person is defiant, or not far enough along in their journey to be open to the Positivity, then one would turn to B. B is an option that is not rude, as it involves one giving oneself to the other person and allowing them an open sounding board. C would be the next and last Method. Method D would never be an option when it comes to

Negative people, unless one is trying to manipulate[51] the Negative person into trusting them, which leads to a whole separate process not discussed in this literary work, and is meant only for skilled healers. Since method D involves losing oneself, which can be dangerous, it would require a Purge afterward. In method B, one can feel confident in the fact that they helped the

---

[51] *Manipulation is never an option for someone Positive, because it involves becoming one with Negative Frequencies, and hiding one's own, with the intention of changing the subject's Wants in one's own interest, instead of inspiring the subject to change of their own interest. Manipulation can lead one down a Negative cycle, wherein one may become even more lost (if they were lost already) due to the Negative outside influence, and by tuning in to Frequencies that may not align with one's True Form. True Positive Change occurs when the subject (on its own) comes to the conclusion of their Wants through Facilitated Learning.*

Negative person, by simply being there for them when they needed to express themselves. This act alone can create Positive vibrations in the other person, and allow for them to release the Negativity they are harboring. However, the release of their Negativity is a possible result, not an expected one! Since each subject (person) is its own System and is under control of the subject's Higher Self, there is always a chance they are far DownStream in a Toxic cycle, and will need much Work in order to again be balanced.

Be cautious with choice B, as it does involve a "Toxic" influence, and therefore allows a chance for you to absorb the Negative Energy the person is emitting (by the ripple effect, and the Railway Model). When utilizing method B, one must learn "Blocking" (discussed later).

Further points on the A, B, C, D Methods

There is one important factor to remember when applying the A, B, C, D Methods: one is never more important, more righteous, or more special than the Negative person. This would create an alternative motive for one's Ego, and can be a form of Negative Energy (as it is not one's purest form).

Food For Thought

The A, B, C, D Methods require one to be clever and quick-witted. One needs to combat Negative Energy before allowing it to build and overflow one's own Positive Energy.

Blocking

Blocking *is intentionally Focusing on not allowing Negativity to affect oneself. One can do this by learning how to tune into oneself* *completely, in order to establish oneself as a separate entity from the environment. In doing so, one would simply Focus on oneself, and "tune in" to one's own Frequencies, thoughts, and Ego. This would also involve selectively tuning in to other Systems surrounding oneself, and would in fact filter the Empathic aspect of oneself. This is known as* ***Introversion***[52].

---

[52] ***Introversion*** *involves blocking oneself off to other systems by utilizing Complete Focus on oneself, one's Thoughts, and one's Ego.*

Food For Thought

After a successful Purge, one can attempt linking one's Frequencies to another's, which involves Focus, and must be without Distraction of the Ego. For a smooth linkage of Frequencies between one entity and another (and for the purposes of this book, strictly human entities), one must also consider that the human whom they are trying to link with is, at the very moment of the linkage, his/her own System, with his/her own thoughts, and therefore vibrating at his/her own Frequencies with his/her own needs. Understanding this about the target of linkage provides one the ability to "tune in" to their Frequencies, and to see their Current Form. This concept is a difficult skill to master, and takes some time and practice, but can be simplified as, "put yourself in their shoes". See how they think, what they are thinking, what their priorities are, what they crave, what they hate, and most importantly, where they are at in their individual journey towards understanding their Higher Self. Beings will link with you when you are Believing a true form of camaraderie and care, and this "linkage" is synonymous to being empathic.

...

...

With these formalities understood, if one is still having difficulties "tuning in" to another's Frequencies, one needs only to establish the value of their own action with oneself, which, for a Positive person, is always the same: because they love that person and wish to help them in any way. The Law of Attraction outlines that you will get a return on your energy output (as Blessings).

5

The Conclusion

Welcome to the Renaissance! Now that you have a plethora of tools to create a better world, where will you begin to spread Positive Change?

The Renaissance: A Time of Discovery for oneself, and for Humanity.

Outline and Synopsis

1. ***Oneself and the Railway Model –*** *"The laughter of children is so pure for a reason."*

(I) Introductory Topics

- *Intro: Purging and Defining Oneself*

- ***The Comfort Zone and Understanding Oneself***
    - *Defining The Comfort Zone and how it contributes to one's image*
- ***A Note Before Entering The Realm of Positivity***
    - *Remember to smile!*
- ***Food For Thought***
    - *The infinite possibilities of "oneself" via Concepts*
    - *Intro to Empath*
- ***The Camera Analogy***

    - *A tool to understand one's operation within the universe, and one's roles and abilities within a given System*
    - *Defining Scope*
- ***Patience***
    - *The practice of waiting to ensure a System is fully understand before acting upon it (to avoid complications in creating Positive Change)*
    - *Defining System*
    - *Intro to the concept of "The Stream"*
- ***Food For Thought***
    - *Negative Effects of Assumptions and Judgments*

(II) The Building Blocks of Creating Positive Change

- *Intro: Keep an open mind, practice Patience and kindness to oneself*

- ***Frequencies***
    - *Opening one's concept of Frequencies*
- ***What are Frequencies?***
    - *Frequencies are interlaced with vibration and energy*
- ***The Relationship of Energy and Frequencies***
    - *"You" are composed of energy*
    - *Energy is always moving and vibrating, and a Frequency is the result*
- ***"Oneself"***
    - *Oneself in the Physical Realm*
    - *Oneself is defined by the energies one is vibrating at, at the time one is measured*
- ***The "Railway Model" and a Central Dogma***

   - *"You" are defined by the Frequencies at which "you" are vibrating*
   - *Intro to the Railway Model – the interaction and connectedness of "you" and the environment*
   - *Intro to String Theory and the connectedness of all things in the Physical Realm*
- ***The Central Dogma of the Railway Model and Further Explanation***
   - *Two Systems of focus that are directly interconnected through energy: the Frequencies of oneself and the environmental Frequencies*
   - *One needs a Proper Railway Model for Positive flow of energy*
   - *A "Toxic" environment leads to Negative Energy flow*

- ***Food For Thought***
  - *One becomes the Frequencies of one's environment*
- ***Respecting One's Railway Model***
  - *Intro to Maslow's Hierarchy of Needs, and the need for Respect*
- ***Maslow's Hierarchy of Needs***
  - *One only Physically needs food, water, and sleep*
  - *A Being, defined*
- ***Food For Thought***
  - *"You" are what "you" eat*
- ***The Railway Model (Further)***
  - *Emphasis that the Railway Model is the interconnectedness between oneself (the self beyond the Physical Realm) and The Body (the link to the Physical Realm)*
  - *The Higher Self, defined*
  - *The Body, defined*

- ***Positive and Negative Energy***
  - *Positive Energy is any combination of Frequencies that lead to happiness, liveliness, inspiration, and is considered a stronger force than "Negativity"*
  - *Negative Energy will be considered anything that is depressing, destructive, or unenjoyable*
  - *Understand that one cannot truly understand the difference until after a Purge is completed*
  - *A Feeling, defined*
- ***Food For Thought***
  - *Positive people are creators, and Negative people are destroyers*

- ***Recap, the Physical Self, and an Introduction to the Basic Principle of Humanity***
    - *Recap the Physical Self ("oneself"/The Body) and its link to surrounding Frequencies*
    - *One can help or hinder the proficiency of one's ability to Manifest one's thoughts into the Physical Realm, by choosing how one treats one's Physical self*
    - *Intro to the concept of The Basic Principle of Humanity*
- ***The Basic Principle of Humanity (BPH)***
    - *The Basic Principle of Humanity, and how all humans are bonded by this principle*

- *Establishing the need to help one another – Humanity is only as strong as its weakest link*
- *Mirror neurons, and the ability of humans to Empathize*

- ***Food For Thought***
    - *As a general rule, if the System as a whole functions Properly, then every component functions Properly, and if every component functions Properly, then the System as a whole functions Properly.*
- ***Depression***
    - *Depression is a state of absolute Negativity, and it can be achievably fixed*
- ***Toxins Re-Established***
    - *Intro to Toxins, and how Negative people are an*

- *Intro: "you" are the "Higher Self"*
- *The concept of utilizing the Realms to create Positive Change*

- ***The Physical Realm, and a Further Introduction to the Higher Realms***
    - *The idea that thoughts have no mass, but are connected to the Physical Realm*
    - *Thoughts have influence over the Frequencies of one's Physical Self*
    - *Intro to Believing*
    - *Intro to Focus*
- ***Food For Thought***
    - *Beliefs change through the process of learning*
    - *Intro to the concept of being "stuck", or unable to grow and*

develop further, A.K.A. "blocked"

- *Limitless possibilities exist when one is open to learning*

- ***Understanding Beauty for Positive Change***
    - *Created Beauty, and Naturally Occurring Beauty*
- ***Food For Thought***
    - *Different Beings have different perceptions, and different ways of reaching the same conclusions*

(IV) Can a Stream Flow with a Dam in its Way?

- *Intro: How does one become "blocked"?*

- ***The Stream Analogy***
    - *Imbalances create blockages*
- ***The Purpose of a Purge***

  - *A Purge is designed to stop or prevent further Imbalances, and to redirect to one's natural flow*
- ***Imbalances***
  - *Imbalances affect the Railway Model*
  - *Imbalances have a cascade effect, and can lead to further Imbalances DownStream*
  - *In the Physical Realm, one can always assume that a System is in fact linked to other Systems somehow, in some way*
- ***Quick Self-Reflection on Imbalances***
  - *What Systems do "you" feel are Imbalanced in your life, and could be causing problems "Downriver", or down the chain of collateral events?*
  - *Have "you" been looking at the DownStream issue (decreasing*

Property values), or the deeper and actual cause of the problem, (the Stream Being blocked)?

2. The Purge – *"Take anything in your life that is Negative, and Physically rid yourself of it—Respectfully, with dignity, pride, and perhaps flowers (if it is a person). Negative Toxins only work to poison your Frequencies, and your Frequencies are your link to change. This is step one in freeing yourself, and is from here on known as purging."*

(I) Purging and The Steps of a Successful Purge

- ***Introduction to Purging***
  - *Purging, defined*
  - *The process of Purging creates a position from which one can initiate Positive Change*
- ***The Steps of a Successful Purge***

  - o *Defining the four steps in a successful Purge: Understanding What a Purge Requires; The Tools of Purging; Implementation of a Successful Purge; Finishing a Purge, and the Results*

- ***Finishing the Concept of The Stream Analogy***
  - o *Understanding there are multiple forces at play, due to the fact there are many interconnected Systems acting upon "the Stream" at the moment of which it is measured*

(II) (A) Understanding What a Purge Requires

- ***Purging, and an Application of Chapter One***
  - o *The concept of oneself as a creator*

  - *Further development on Beings of Consciousness*
  - *Intro to the Natural Peace of Energy*
- ***Toxic Entities and Purging***
  - *Purging involves ridding oneself of every genre of Negativity*
- ***Purging Negative People***
  - *Purging Negative People is okay*
  - *One must first help oneself in order to be able to help others*
- ***Food For Thought***
  - *As a general rule, Negative people lead to Negative results, and Positive people lead to Positive results.*
- ***Understanding and Identifying a Toxic Environment***
  - *One must understand what "Toxins" mean to oneself,*

which will be more clear after defining one's Wants

- ***Excuses***
    - *Excuses, defined*
    - *The concept of awareness of one's own thinking patterns and how one could be stuck in a Negative thought pattern*
    - *Introducing the concept of The Box and how one must change one's thoughts to Manifest Positive Change*
- ***The Box***
    - *Understanding that The Box is a self-conceived concept of oneself, defined by limitations*
- ***Excuses (Continued)***
    - *Common societal excuses for why something cannot be done are symptoms of confinement to The Box, i.e. "I am too old".*

   - *Practice learning, establish Wants, and Purge correctly to liberate oneself from The Box*
- ***Food For Thought***
   - *Creativity can solve any problem*
- ***Understanding How to Become Limitless, In Order to Have Successful Purge***
   - *One is limitless, due to the fact one is born a Tabula Rasa, and can learn information*
   - *Classical Conditioning*
   - *One can always create or destroy one's own Box*
- ***Food For Thought***
   - *One can always create or destroy one's own Box*
- ***The Impossible Has Been Proven Possible***

- *In history, Beings have proven the "impossible" to be possible (The Wright Brothers, Lance Armstrong, etc.)*

- ***Irrational Limitations and Utilizing Complete Focus***
    - *Diction and thought patterns determine what "you" are capable of*
    - *Words are very powerful forms of energy and Manifestation*
    - *You are what you Focus on*
- ***Food For Thought***
    - *Using one's voice is a way of transmitting Frequencies*
- ***Concluding "Understanding What a Purge Requires" and Intentional Forgetting***
    - *The concept of Intentionally Forgetting in order to redirect*

one's energy towards a Purge and therefore Positive Change

(III) (B) The Tools of Purging

- ***Wants***
    - *What one Wants is a Focus of one's thoughts*
    - *Introducing the concept of self-fulfilling prophecies*
- ***Establishing Complete Focus***
    - *Wants are a Manifestation and a pas de deux of one's Physical and Higher Selves*
    - *Establishing Complete Focus is essential in Maintaining one's Wants*
    - *Distractions are anything that do not have to do with one's Wants*
- ***Ego***
    - *Wants are considered to stem from the Ego*

  - *The Ego is a Distraction to creating Positive Change*
  - *Introduction to True Form and Current Form, and how each can be understood to check one's own Ego and help others*
  - *Everything in moderation*
- ***Controlling One's Emotions in Order to Contain the Ego***
  - *Emotions, defined*
  - *One must first understand oneself in order to control one's Ego*
  - *We have the choice of how to respond to Emotions*
  - *Example of Sally Sue and Billy Bob*
- ***Complete Focus Continued***
  - *One must always start by Maintaining oneself. Everything else (the ability of Complete*

Focus, and other tools of Positive Change) will then follow suit

- *Intro to Gauges*

- ***Gauges, and a Brief Introduction to Remaining "In the Moment"***
    - *Gauges allow one a snapshot of a particular moment*
    - *Gauges are a tool for understanding how to remain in the moment*
    - *One must understand for any given situation, one can only control the proceeding actions, and cannot change past happenings*
- ***The Distractions of Past and Future***
    - *When Maintaining Complete Focus, past and future can be distracting*

  - *Past and future can cause one to misread Gauges, which is why Complete Focus is critical in remaining in the moment*
  - *Gauges and its link to athletics*
  - *Worry and Anxiety are Distractions that bring one to the past or future*
- ***Complete Focus Finalized***
  - *Complete Focus is without Distraction*
  - *One cannot fully utilize Complete Focus without first being in balance. One will not be in balance until after a successful Purge*
  - *Anything that is not Priority is irrelevant when practicing Complete Focus*
- ***Ridding Oneself of Distractions***

- *This process may cause internal conflict, between the Logic of the Higher Self, and Emotions felt Heart (Physical Self)*
- *Purging gets rid of Toxic Distractions, and assists in achieving Complete Focus*

- ***Defining One's Individual Wants***
  - *There are NO excuses when defining one's Wants, and no time frames*
  - *What does one Want to do with one's life next?*
  - *Allow oneself to fail hundreds of times when defining one's Wants – failure is the greatest teacher*
  - *Understanding Toxins in disguise*

    - *Using the "test" method to sift through and identify Toxins*
- ***Defining One's Individual Values***
    - *After one establishes Wants, one can begin to define what Values one needs in order to Manifest those particular Wants*
    - *Best Possible Fireman example*
- ***Food For Thought***
    - *One should not fully understand what it takes to become the best possible fireman*
    - *when first setting out for that particular want*
    - *It is about the journey, not the destination*
- ***Being Clear With One's Wants, In ALL Moments: A Practice of Complete Focus***

  - *A key to the Purge process is being clear with one's Wants in every moment*
  - *Consistency is another tool of a successful Purge*
- ***Finding Balance During a Purge***
  - *One needs balance in order for alignment, and in order for a successful Purge*
  - *A Purge is a process of finding and creating balance*
  - *Extremes must be taken sometimes in finding a complete balance*

(IV) (C) Implementing a Successful Purge

- ***Recap of Concepts So Far, and Initiating the Purge***
  - *Toxins are Negative, and therefore oppose that which is Positive*

  - *The first step: understanding Toxins*
  - *Thoughts stem from the Higher Self*
  - *The Physical Self is guided by thoughts*
- ***Food For Thought***
  - *Negativity has a "weighing down" effect, and Positivity has a very light effect*
- ***One Will Manifest Changes to the Physical Realm Through Beliefs***
  - *Believing something is a matter of choice. One must choose to Focus on and Believe an entity for it to be Manifested in the Physical Realm*
- ***Express Oneself!***
  - *Keep consistent expression of and dedication to your Purge*

practice, amongst all Systems in one's immediate connectedness
- o *The Results Equation illustrates the dogma that the Factors put in to a System = the Results one gets out*

- ***A Moment to Celebrate***
  - o *Understanding one is a creator of one's own life*
  - o *After a Purge, one will begin to create one's new life*
- ***Time to Purge***
  - o *A Purge is a time of great challenge*
  - o *Understand how to use all of the tools*
  - o *A Purge will not be a comfortable thing*
- ***Food For Thought***
  - o *Knowledge is Power*

(V) (D) Finishing a Purge, and the Results

- ***Re-Introduction to Complete Focus***
    - *Evolution of the concepts of Manifestation and the power of thoughts*
    - *Complete Focus of Positivity and Intentional Forgetting of Negative influences*
    - *One will still need to refine one's distinction between Positive and Negative, and trial and error will be the best way to learn*
- ***Characteristics of The End of a Purge***
    - *Change involves physical and mental adaptation*
    - *When one accepts change, one allows oneself to evolve, and it is though Respect that these changes remain in power*
- ***Time***

- o *Time is an incremental reference of change in the Physical Realm*
- ***Food For Thought***
    - o *One must beware that one is not working to preserve a Negative cycle*
- ***Following a Purge***
    - o *A successful Purge will entail that one has now become balanced, has stopped a Toxic System (a Negative cycle), has experienced much change, uncertainty, and inner tension, no longer has any Negative influences in their immediate Railway Model, and is now at a place where one can begin again to develop and create an entirely new System, with entirely new Frequencies,*

without the ability of returning to the old Frequencies (or the old Comfort Zone).

- ***Purge Completed***
    - *A Negative cycle has ended*
    - *Zero is a point from which one can create a new cycle*

3. Recreating Oneself and Quantifying Results – *"I stopped looking [for Positivity] and let it find me. As for [Negativity], it's part of life, but has no place in my life. Love all, and let those that hate, keep hating. Decide from this moment what kind of person you Want to be, and live accordingly." - Unknown*

(I) "The After Purge"

- ***Introduction to "The After Purge"***
    - *Celebrate one's successful Purge!*
- ***Things to Note in "The After Purge"***

    - *One will now be at Zero (a new place) with no chance of returning*
    - *A recognition that as humans we are very young and have much to learn*
- ***Extended Durations of Toxicity and the Rehabilitation Period***
    - *One must allow oneself a Rehabilitation Period when returning to Zero*
- ***The Result of a Purge, and Shifting Focus to Recreating Oneself***
    - *Notes on the connection between the Higher Self and the Physical Self*
    - *Emphasis on the concept that the efficiency of one's Railway Model is dependent upon the condition that it is kept*

(II) Recreating Oneself

    - *Intro: Just do it, and DO NOT give yourself time to think. There can be no Distractions.*
- ***Introduction to Recreating Oneself***
    - *After a Negative Cycle, one must balance oneself out (The Rehabilitation Period)*
    - *Building a strong foundation will depend on the work put in to its layers*
- ***Breaking Things Down, to Build Them Back Up: Start With a Strong Foundation***
    - *A strong foundation is the outcome of layers and layers of Positive results*
- ***The Results Equation***
    - *Results = Focus x Energy x Time*

- *Results are the outcome of a System at the time at which it is measured*
- *The Results equation is a conceptual tool, with practical uses*

- ***Food For Thought***
  - *When applying the concepts of The Results Equation, ensure to balance quality over quantity*
- ***The Factors and The Uses of The Results Equation***
  - *R = (F [+ or -] Q) x (E [+ or -] Q) x (T [+ or -] Q)*
  - *Allows one to determine the quality and Result for given inputs of F, E, and T*
- ***The Results Equation Continued***
  - *A successful Purge takes maximum amount of inputs*
- ***Food For Thought***

    - *The Results Equation always holds true*
    - *Any System always Maintains its strongest roots the longest before inevitably dwindling (if no Work is put in)*
- ***Maintenance of a System***
    - *The combination of Complete Focus, Patience, Work, and Respect is Maintenance*
    - *Further into Entropy*
- ***Elysium***
    - *Introducing a System of Beauty: Elysium*
- ***How to Become a Diamond***
    - *Diamonds start from fragile substances, and are formed under immense amounts of pressure. They are incredibly strong*

- *To become a Diamond, one must withstand a great deal of pressure (or challenges)*
- *Definition of Failure – the greatest achievement of all*

- ***Food For Thought***
  - *Diamonds figuratively signify strong bonds*

(III) Setting the System – *"Doing all the right things"*

- ***Finding Balance, and a Conclusion of Concepts***
  - *Recapping concepts of balance and achieving Elysium*
- ***Creating Positive Change***
  - *Manifestation of the vision is the product of consistency*
  - *Attuning oneself to one's vision*
  - *Positive Change involves steadfast Intuition*

- ***Intuition***
    - *The ability to freely Manifest one's thoughts into the Physical Realm*
- ***The Four Choices of Any System***
    - *Once a Properly flowing System is established, it need only be Maintained in order to prevent entropy*
    - *One can Maintain a System, grow a System, destroy a System, or remove awareness with Intentional Forgetting*

4. Positive Change on a Massive Scale *– We will stray far from the traditional yin-and-yang (+ and -) concept of Energy, to speak more of its Manifestation through human nature. Step into an amazing vision of a world where Positive Energy thrives!*

(I) The Positive Movement *– "…We (Humankind, all bonded by the Basic Principle* *of Humanity) are only as strong as our weakest link, and even our weakest links thrive in happiness."*

- *Intro: Positive Overhaul*

- ***The Positive Movement***
  - *The Positive Movement is a formal coterie that any Positive Being intrinsically belongs to*
- ***The Law of Attraction***
  - *Positive Frequencies lead to Positive Frequencies, and Negative Frequencies lead to Negative Frequencies*
  - *The energy one is creating with Positive thoughts will Manifest in the Physical Realm in Positive ways, i.e. in the form of Blessings*

(II) Positive Change on a Massive Scale, and Becoming a Positive Healer

- *Intro: "The planet does not need more successful people. The planet desperately needs more peacemakers, Healers, restorers, storytellers, and lovers of all kinds." – Dalai Lama*

- ***Energy Work***
    - *Energy Work operates using different Frequencies and can influence Physical structures*
- ***An Energist***
    - *Energists are Healers, operating in their True Form, free from their Ego*
- ***Using Frequencies In Order to Influence the Physical Self***
    - *Frequencies can be used to manipulate the Physical Self in varying ways*

- ***The Application of Positive Frequencies in Creating Positive Change***
    - *When creating Positive Change, Positive Energy needs to exist solely as a crushing force to Negativity*

(III) Further Necessary Tools for Positive Change

- ***Introduction to Changing the Flow of a Moment***
    - *Once one has become practiced enough at creating Positive Change for oneself, one will also be able to change the flow of a moment while interacting with others*
    - *Miscommunications can result in a Negative cycle*
- ***Encountering a Deeply Negative Being***

- *When a Positive Being encounters a truly Negative Being, a sort of combat of Frequencies begins*
- *Intro to the A, B, C, D Methods of Positive Change*

- ***The A, B, C, D Methods of Creating Positive Change: When Dealing With Negative Entities***
  - *There are four possible courses of action when Positive and Negative Beings encounter one another*
  - *A) Attempt to influence the subject Positively through Facilitated Learning, B) Allow the subject to express oneself, and do nothing at all, C) Purge, or D) "Tune in" to the Frequencies at which the subject is operating*

- *Quick explanation of method D*
- *Stay Humble*

- ***Food For Thought***
  - *The A, B, C, D Methods require one to be "clever" and "quick witted": one needs to combat Negative Energy before allowing it to build and overflow one's Positive Energy*
- ***Blocking***
  - *Intentionally Focusing on not allowing Negativity to affect oneself*
  - *Introversion, defined*
- ***Food For Thought***
  - *Linking one's Frequencies to another is a matter of Focus*
  - *Performing this linkage is a sign of empathy*
  - *Positive people find value in "tuning in" to others because*

they love that person, and wish to help them in any way

- *The Law of Attraction will return the energy output to you with Blessings*

5. *The Conclusion* – *Welcome to the Renaissance! Now that you have a plethora of tools to create a better world, what will you do?*

Glossary

(Important Concepts and Definitions)

The A, B, C, D Methods of Creating Positive Change – *A four choice method when dealing with Negative entities. A) Attempt to influence the subject Positively through facilitated learning, B) Allow the subject to express themselves, and do nothing at all, C) Purge, or D) "Tune in" to the frequencies at which the subject is operating. (Pg. 159)*

The Basic Principle of Humanity (BPH) – *The binding fact amongst all humans: the idea that control of one's Physical Self has been bestowed to each of us (including you, the reader of this work) in some way (however one believes—through fate, some type of God, the Universe, spontaneity, just plain no reason, etc.). It entails that by Respecting oneself, one is also Respecting others, and will never be projecting a Negative Frequency. It also entails that since, as Physical Beings, we are all a part of the same System (the Physical Realm), we therefore can be considered separate entities of the same whole. (Pg. 38)*

Beauty – *Beauty can be seen in two ways: Beauty that is created, and Beauty that is found naturally occurring. It is* *always the observable result of Positive Energy. One can quantify Beauty to be the result of a maximum amount of Positive Work put in for a given unit of Time. (Pg. 50)*

A Being – *A Being is classified as something that has an ability to create change in their environment through manipulation of a physical form, involving a form of consciousness. (Pg. 30)*

Beings of Consciousness – *Beings of Consciousness are those that have the capability of becoming highly self-aware. Humans are capable of analytical thought, thus giving us "consciousness", the ability of Facilitated Learning, and also separating us from entities incapable of self-awareness. Consciousness grants us the ability of creating Positive Change with Proper understanding. (Pg. 43)*

Believing – *1) A conscious choice, supported by thought, to choose to Focus on something that is understood to be "true" in the reality of the Being that has perceived it. Thus, a Belief is defined as a consistent stake in a particular set of thoughts. 2) An agreement between the Higher Self and the Physical Self, allowing for a smooth flow of energy, and balance. Beliefs change when more information is accrued (Learning) for the refinement of that particular Belief. Believing something isn't magic.*

Believing something is a concrete matter of choice, strengthened by thought patterns, and changed through Discipline. (Pg. 49)

Blessings – *Unforeseen bits of good-luck, or Positivity, in any form; one result of successfully completing a Purge. (Pg. 150)*

Blocking – *Intentionally Focusing on not allowing Negativity to affect oneself. One can do this by learning how to tune into oneself completely, in order to establish oneself as a separate entity from the environment. (Pg. 163)*

The Body – *The conduit between thoughts of the Higher Realms and the manifestation of those thoughts in the Physical Realm. It is one's link to the Physical Realm, and the defining aspect in the ability for one to have a Railway Model. The Body is the "camera" in The Camera Analogy. (Pg. 33)*

The Box – *A figurative representation of mental obstacles. The walls, or limitations, of the mental structure one has set oneself up with, and is strengthened by one's own thinking patterns. The Box is a visual example of one Being "trapped" in a box by the "thoughts" that one has put up as obstacles to change. It is the biggest obstacle when it comes to Positive Change. (Pg. 73)*

The Camera Analogy – *A useful tool to understand the universe, and one's roles and abilities within a given system that one is interacting with. One is to think of oneself as a camera, and a System as what one sees through the camera Scope. Just as a camera has the ability to zoom in and out, far and wide, so too does one have the ability to see any System that falls within, but not outside of, one's adjustable camera Scope for the given moment that one is making an observation. (Pg. 17)*

Concept – *A compilation of thoughts that create a completed idea. Concepts vary in strength by the amount and legitimacy of the supporting thoughts creating it. (Pg. 15)*

Comfort Zone – *The mental limitations or boundaries one places upon oneself; commonly known as one's "image". The Comfort Zone is used to describe the walls of The Box. (Pg. 14)*

Complete Focus – *A state of Discipline in which there are no Distractions from the entity one is choosing to Believe at the time of which the Focus is being applied. Complete Focus is the absolute key in manifesting Positive Change, and the final tool one needs in order to complete a Purge. Complete Focus entails one still allows Positive influence,* *but ignores (Intentionally Forgets) Negative influences. (Pg. 74)*

Current Form – *Current Form is the alternative form to True Form. It is the form one is currently Manifesting. Current Form is determined by the Being's knowledge, how that Being's Railway Model is functioning, and that Being's understanding of other concepts in this book. For example, perhaps the other Being is unaware of its Ego (leading to some psychological "disorders", i.e. Narcissism), or its own consciousness, and therefore that Being will manifest in its Current – unbalanced – Form. (Pg. 84)*

Depression – *A state of absolute Negativity. One that is depressed may feel as though they are "trapped", "limited", "unable to..." or any other variations of this sort. The reversal of this overflow of Negative Energy is the main application of a Purge. (Pg. 41)*

Diamond – *An immensely strong physical material, and is the end result of a lengthy passage of time, involving immense amounts of pressure. To metaphorically become a Diamond, one must apply an extreme amount of pressure in order to strengthen the layers of bonds that hold oneself together. (Pg. 135)*

Discipline – *The controlled practice of Focus. (Pg. 73)*

Distractions – *Anything that do not have to do with one's Wants; entities considered to derail one's Focus. (Pg. 82)*

DownStream – *The resulting effects; the effects of an Imbalance that can be traced back a few connections in a cycle, or that can be thought of as a result of one or more Imbalances. (Pg. 60)*

The Ego – *A non-Physical location from which a Want (not a need) of one's individual Physical Self stems from. The Ego is what creates oneself to be separate from the Universe. Without an Ego, one simply becomes that which is: the energy that is omnipresent. Control of the Ego provides clarity, greater intuition, the ability of empathy, awareness, and Complete Focus. (Pg. 83)*

Elysium – *Elysium is a System in which Positive Energy is so foundationally strong, and properly Maintained, that no Negative Energy can influence it. A System of Created Beauty and pure Positivity. (Pg. 135)*

Empath – *Someone who can understand that there is an infinite amount of possibilities of "oneself". An Empath is able to look at another human and understand exactly what they consider to be their individual "oneself". Empaths are the creators of Positive Change. (Pg. 15)*

Emotions *– An emotion is the perception that a Being identifies with a particular experienced Feeling (Pg. 85)*

Energist *– A Being that understands the flow of energy through a System, and how to guide energy for Positive healing purposes. A true Energist is also a skilled Empath, and has the ability to assist in healing the Railway Models of others. (Pg. 153)*

Energy Work *– The utilization of different vibrations in order to influence physical structures; comes in many forms: Meditation, Electric Stimulation Therapy, Reiki, Tai Chi, Massaging, Technology, etc. (Pg. 152)*

Entropy *–The Universe has a tendency towards disorder; in the occasion of Positive Change, Entropy defines the principle that observes that the System will collapse if no Proper maintenance or Work is performed upon it (Pg 134)*

***Excuses* –** *Any reason one cannot be able to do something. Excuses are always invalid, and a sign of being "stuck". (Pg. 72)*

Facilitated Learning *– The purposeful passing of information from one Being to another. This can occur through many venues (some of which include, but are not* *limited to, tactile, auditory, and visual segues of information transfer). (Pg. 43)*

Failure *– With the correct Positive mindset, Failure is by far the greatest achievement of all. Failure implies one has attempted something, has not succeeded yet, and now understands something new about the venture they are trying, which in turn becomes a step further in the process of creating that venture. True failure is to repeat the same mistakes, or never to try at all. (Pg. 137)*

Feeling *– The process of experiencing a Frequency. When one experiences a feeling, one is given a choice: to manifest that particular Frequency, to "tune in" to another Frequency, or to do nothing. (Pg. 35)*

Focus *– The act of choosing one entity to think about, and Maintaining a directed thought flow towards conceptions of that one entity, for any duration of time, in order to Manifest it. Focus is a method of controlling energy flow. (Pg. 49)*

Frequency *– The result of the energy and vibration of a particular System at the time of which it is observed. It is the term used to describe a particular energy and how it travels. At any given point, a particular System is defined by the Frequency to which it is vibrating at the time of which it is observed. (Pg. 23)*

Gauges *– Gauges are the truest and purest form of understanding a current situation, without the hinderance of the past, the future, the Ego, or any other Distractions; they are a snapshot of a current moment, in that moment's purest form. Gauges involve a constant assimilation with oneself to the surrounding Frequencies, and an openness to go with the flow, accept change, and appreciate what it is that one has been presented with, in real time. One can always utilize Gauges for Positive Change, in any given situation, at any moment, for there is always Beauty to be found in any particular moment, whether one presently realizes it or not. (Pg. 89)*

Heart *– The Physical entity of emotional feeling. If in a state of Imbalance, one's Heart may conflict with one's Logic. (Pg. 94)*

The Higher Realms *– One of two realms. One's Being, one's Higher Self, the visual and limitless realms (not limited by the laws of the Physical Realm), and the origin of the higher forces (all possible Systems of the Physical Realm). The Realm of "Creators" and everything else. For simplicity: thoughts, Beings, and everything else. (Pg. 44)*

Higher Self *– In The Camera Analogy, it is the entity that utilizes the camera itself (that which is doing the viewing).*

It is above the camera and the Scope in terms of existence. (Pg. 33)

<u>Imbalances</u> – *Anything that creates blockages in a naturally flowing System. (Pg. 58)*

<u>Intentional Forgetting</u> – *The practice of forcing oneself to Focus on Positive entities (implemented after one has recognized Negative thought patterns). This grants the awareness of one's current mental direction, the ability to refine and tailor one's thought patterns (or to Completely Focus them), and the absolute manifestation of all things pertaining to one's Wants (a key practice in order to complete a successful Purge). This process also ensures one will no longer waste one's energy (since it entails eliminating "limiting" entities in one's thoughts) wherein that energy can then be redirected towards the goal of a successful Purge and Complete Focus. (Pg. 78)*

<u>Introversion</u> – *Blocking oneself off to other Systems by utilizing Complete Focus on oneself, one's Thoughts, and one's Ego. (Pg. 164)*

<u>Intuition</u> – *One's ability to freely manifest one's thoughts (or energies) into the Physical Realm, without hesitation, tension, or any setbacks. Intuition is a sign of balance, and implies one understands what one knows, and what one doesn't know (utilization of one's Gauges), for any given* *moment. This allows for Proper and immediate decision making-to-action. (Pg. 143)*

The Law of Attraction – *The Dogma that Positivity attracts Positivity, and Negativity attracts Negativity. The energy one is creating with Positive thoughts will manifest in very Positive ways in the Physical Realm, therefore one will naturally "attract" new Positive opportunities to oneself, due to the Properties of the Physical Realm, Elysium, and the smoothly flowing Railway Model. (Pg. 149)*

Little Albert Experiment – *Humans learn through experience. They are not preprogrammed, but are constantly adapting and changing (learning) through new life experiences and encounters. The new discoveries regarding the epigenome and its ability to influence one's genetic coding, neuroplasticity, and the experiment of Little Albert, which proved humans are born a "tabula rasa", all support the strong opinion that nurture is the primary segue for learning and change. But these examples do not stand alone: my own and others' experiences and encounters with feeling it to be "impossible" to break out of what was thought to be a hopeless and preset "box" (The Comfort Zone), yet the fact we have all successfully created new lives for ourselves, despite our thoughts that "nature" was our infinite downfall, further prove the* *power and influence of nurture. This is commonly known as the nature versus nurture argument. (Pg. 22)*

Logic – *Calculations made in response to the Physical Realm that are concurrent with one's priorities. If in a state of Imbalance, one's Logic may conflict with one's Heart. (Pg. 94)*

Maintenance – *The combination of Complete Focus, Patience, Work, and Respect. Requires enough Work to prevent Entropy. It can also be quantified mathematically by The Results Equation. (Pg. 133)*

Manifest – *The transference of thoughts, dreams, desires, etc. (ultimately, the Positivity one wishes to create) from the Higher Realms into the Physical Realm. (Pg. 36)*

The Natural Peace of Energy – *The Natural Peace of Energy (or the Natural Flow of Energy) describes a state of balance in which energy is not hindered by blockages, and can freely flow through entities of the Physical Realm. (Pg. 69)*

Negative Energy (Negativity) – *Any combination of Frequencies that lead to depression, manipulation, selfishness, or lack of enjoyment. Negativity has a "weighed down" effect, and is destructive. (Pg. 35)*

Patience *– A practice of ensuring a System is fully understood before acting upon it, without time constraints. (Pg. 18)*

The Physical Realm *– One of two Realms. The Physical, the tangible; things with mass, things with energy. Things that are limited by the Physical laws of nature, and bound by Time. For simplicity: it contains The Body, and things bound by Time. (Pg. 44)*

Positive Energy (Positivity) *– Any Frequencies known to exist on the higher wavelengths. These Frequencies lead to feelings such as happiness, liveliness, and inspiration, and are contrasting forces to Negativity. Positivity has a light effect, and is constructive. (Pg. 34)*

Priority *– Priority is a logical listing of what is most important, or what one Wants the most, and, in the occasion that there is more than one Want, is arranged in order of importance to that individual Being. (Pg. 93)*

Product *– A Physical, tangible result of a System. (Pg. 130)*

Proper *– Proper is used to describe a well-Maintained connection, something respected, or in the occasion of a Proper environment, an environment in which Positive Change can flourish. Used to describe the ideal state of one's Railway Model. (Pg. 27)*

Purging *– Purging is the Process of removing Toxins, and therefore Imbalances, from one's System, in order to create a position for one to initiate Positive Change. There are four steps or phases that one will understand and undergo when beginning and completing a Purge: (A) Understanding What a Purge Requires, (B) The Tools of Purging, (C) Implementing a Successful Purge, (D) Finishing a Purge, and the Results.*

Railway Model *– The "Railway Model" is a tool used to simplify and describe the flow of energy between "oneself" and the environment, and highlights their interconnectedness. It can also describe the interconnectedness between oneself and the body. A Proper flowing "Railway Model" is key to deeming a successfully completed Purge. (Pg. 25)*

Rehabilitation Period *– An allotted amount of time in which one allows oneself (through Work) to return to Zero. The success of the Rehabilitation Period will be defined by The Results Equation. (Pg. 125)*

Respect *– The practice of understanding the power and ability something has; putting Work and Time into Proper Understanding of something. (Pg. 29)*

Respecting *– The act of giving Energy to something in order for its Positive development. (Pg. 29)*

Result(s) – *The outcome of a System at the time at which it is measured. (Pg. 130)*

The Results Equation – *A central dogma: the factors one puts in to a particular System is directly proportional to the Results one will get out. R = (F [+ or -] Q) x (E [+ or -] Q) x (T [+ or -] Q), where R is Results, F is Focus, E is Energy, T is Time, and Q is the Quality of those variables. (Pg. 131)*

Scope – *The Scope is a piece of The Camera Analogy that describes the observable area for which information is taken by the camera. (Pg. 17)*

The Stream Analogy – *The Stream Analogy is used to simplify a greater concept of how the flow of energy in our Universe operates, and to illustrate how to self-correct broken or blocked flows. (Pg. 56)*

String Theory – *The connection of all matter and concepts to one another in the Physical Realm. It is a vast topic, but in a general sense, it means everything is connected in some way, like strings of varying strength. The example of the Stream, in Section (IV) of Chapter I "Can a Stream Flow with a Dam in its Way?" will help to elucidate upon the interconnectedness of things of matter. (Pg. 25)*

System – *All the connected entities of anything that exists in the Physical Realm. There is an infinite amount of overlapping Systems that complete a "Stream" of energy. In The Camera Analogy, a System is what one sees through the camera Scope. (Pg. 18)*

Time – *A reference of change in the Physical Realm. A moment in Time is defined by the Frequencies that were occurring at the point during which it was measured. (Pg. 115)*

Toxins – *Toxins are defined as people and things that have a Negative, or Imbalancing influence. They can result in a hindrance of one's ability to have a Proper functioning Railway Model. If one is feeling depressed, trapped, or lost, one can safely assume there is some source of Toxins holding one back from growing, or feeling at peace. (Pg. 28)*

True Form – *The True Form of other Beings is the form at which they manifest when they are balanced. It is the form at which one is completely Positive and in touch with one's Wants. (Pg. 84)*

Values – *The particular skillset that is required for Manifesting a Want, as well as the substance (qualities,* *vision, etc.) that makes up the Want one is establishing. (Pg. 99)*

Wants – *Unconsciously or consciously defined priorities that stem from one's Ego. Answering the question, "What does one truly Want?" allows oneself to begin to create boundaries for their individual Purge. The process of establishing one's Wants is a practice of Complete Focus. (Pg. 74)*

Work – *Focus times Energy. (Pg. 134)*

Worry – *An emotional state of Imbalance in which one is projecting one's Higher Self to the past or the future, and is therefore missing the current moment. (Pg. 91)*

Zero – *The representation of the point in time from which one can begin to create a new cycle. A place of infinite possibilities. (Pg. 117)*

About Author Nicholas Gambacini

From Decorated Athlete, to Biologist, to Positive Public Figure, to Exercise Therapist, Nicholas invites us all to be eager in our own quests for understanding and experience. "My work is one of selflessness and little to no return: wherein most authors work to pay their own bills, I work to pay the bills of society, and collection is constantly due. I, and my following, believe in the eradication of depression, environmental issues, and more. This is the only debt I pay, the debt of a society that has far more potential than it is currently given credit for."

Having been an author since the age of 16, today his works vary from a wide range. As a coach and training athlete, Nicholas sets out to provide the means of fostering positive growth for younger athletes; as an Exercise Therapist, he jumps into the trenches to pull others through the day-to-day grind that can be seemingly impossible fitness goals; and as an individual, he finds joy in the beauty of art, physics, and positive community growth. "I have been a part of many societal roles, and the one on the pedestal is no different than the one below. The truth is, they are all the same. If I can exemplify the power that dirt can have, if I can do this from nothing, then others can, especially those that are the powerful influences for generations to come." Nicholas constantly emphasizes the importance of community, the power of the individual, and the self-worth that hard work invariably holds.

Finnerman's Authors Association

**Established 2018*

Dear Nicholas Gambacini,

On behalf of the Finnerman's Authors Association, I want to thank you for enrolling your literary work, "Creating Positive Change", in our Charitable Donors Program. By enrolling in this program you have agreed to donate a percentage of your proceeds to a charity of your choice. Proudly display our Charitable Donors logo on the cover, and within the text of "Creating Positive Change", so that your readers will know they have not only purchased a book worthy of the Finnerman's Authors Association, but have also helped fund a charitable cause, and therefore have become a patron of humanity.

Thank you for helping create a more positive world for generations to come.

Sincerely,

Janet L Hughes

"Janet Hughes"

Finnerman's Authors Association Chair Head

Made in the USA
Monee, IL
04 November 2026

67378309R00134